Chaos in Partial Differential Equations

Graduate Series in Analysis

Chaos in Partial Differential Equations

Y. Charles Li

University of Missouri, Columbia

 International Press

Library of Congress Cataloging-in-Publication Data

Li, Y. Charles,
       Chaos in Partial Differential Equations / Y. Charles Li.
       129p. cm. (Graduate Series in Analysis, v. 2)
       Includes bibliographical references and index.
       ISBN 1-57146-151-5
       1. Partial Differential Equations 2. Chaotic Dynamics I. Title
QA 614.8

Typeset using the LaTeX system
Printed in the U.S.A. on acid-free paper

Contents

Preface

The area: Chaos in Partial Differential Equations, is at its fast developing stage. Notable results have been obtained in recent years. The present book aims at an overall survey on the existing results. On the other hand, we shall try to make the presentations introductory, so that beginners can benefit more from the book.

It is well-known that the theory of chaos in finite-dimensional dynamical systems has been well-developed. That includes both discrete maps and systems of ordinary differential equations. Such a theory has produced important mathematical theorems and led to important applications in physics, chemistry, biology, and engineering etc.. For a long period of time, there was no theory on chaos in partial differential equations. On the other hand, the demand for such a theory is much stronger than for finite-dimensional systems. Mathematically, studies on infinite-dimensional systems pose much more challenging problems. For example, as phase spaces, Banach spaces possess much more structures than Euclidean spaces. In terms of applications, most of important natural phenomena are described by partial differential equations – nonlinear wave equations, Maxwell equations, Yang-Mills equations, and Navier-Stokes equations, to name a few. Recently, the author and collaborators have established a systematic theory on chaos in nonlinear wave equations.

Nonlinear wave equations are the most important class of equations in natural sciences. They describe a wide spectrum of phenomena – motion of plasma, nonlinear optics (laser), water waves, vortex motion, to name a few. Among these nonlinear wave equations, there is a class of equations called soliton equations. This class of equations describes a variety of phenomena. In particular, the same soliton equation describes several different phenomena. Mathematical theories on soliton equations have been well developed. Their Cauchy problems are completely solved through inverse scattering transforms. Soliton equations are integrable Hamiltonian partial differential equations which are the natural counterparts of finite-dimensional integrable Hamiltonian systems. We have established a standard program for proving the existence of chaos in perturbed soliton equations, with the machineries: 1. Darboux transformations for soliton equations, 2. isospectral theory for soliton equations under periodic boundary condition, 3. persistence of invariant manifolds and Fenichel fibers, 4. Melnikov analysis, 5. Smale horseshoes and symbolic dynamics, 6. shadowing lemma and symbolic dynamics.

The most important implication of the theory on chaos in partial differential equations in theoretical physics will be on the study of turbulence. For that goal, we chose the 2D Navier-Stokes equations under periodic boundary conditions to begin a dynamical system study on 2D turbulence. Since they possess Lax pair and Darboux transformation, the 2D Euler equations are the starting point for an analytical study. The high Reynolds number 2D Navier-Stokes equations are viewed as a singular perturbation of the 2D Euler equations through the perturbation parameter $\epsilon = 1/Re$ which is the inverse of the Reynolds number.

Our focus will be on nonlinear wave equations. New results on shadowing lemma and novel results related to Euler equations of inviscid fluids will also be presented. The chapters on figure-eight structures and Melnikov vectors are written in great details. The readers can learn these machineries without resorting to other references. In other chapters, details of proofs are often omitted. Chapters 3 to 7 illustrate how to prove the existence of chaos in perturbed soliton equations. Chapter 9 contains the most recent results on Lax pair structures of Euler equations of inviscid fluids. In chapter 12, we give brief comments on other related topics.

The monograph will be of interest to researchers in mathematics, physics, engineering, chemistry, biology, and science in general. Researchers who are interested in chaos in high dimensions, will find the book of particularly valuable. The book is also accessible to graduate students, and can be taken as a textbook for advanced graduate courses.

I started writing this book in 1997 when I was at MIT. This project continued at Institute for Advanced Study during the year 1998-1999, and at University of Missouri - Columbia since 1999. In the Fall of 2001, I started to rewrite from the old manuscript. Most of the work was done in the summer of 2002. The work was partially supported by an AMS centennial fellowship in 1998, and a Guggenheim fellowship in 1999.

Finally, I would like to thank my wife Sherry and my son Brandon for their strong support and appreciation.

CHAPTER 1

General Setup and Concepts

We are mainly concerned with the Cauchy problems of partial differential equations, and view them as defining flows in certain Banach spaces. Unlike the Euclidean space $\mathbb{R}^n$, such Banach spaces admit a variety of norms which make the structures in infinite dimensional dynamical systems more abundant. The main difficulty in studying infinite dimensional dynamical systems often comes from the fact that the evolution operators for the partial differential equations are usually at best C^0 in time, in contrast to finite dimensional dynamical systems where the evolution operators are C^1 smooth in time. The well-known concepts for finite dimensional dynamical systems can be generalized to infinite dimensional dynamical systems, and this is the main task of this chapter.

1.1. Cauchy Problems of Partial Differential Equations

The types of evolution equations studied in this book can be casted into the general form,

$$(1.1) \qquad \partial_t Q = G(Q, \partial_x Q, \ldots, \partial_x^\ell Q) \ ,$$

where $t \in \mathbb{R}^1$ (time), $x = (x_1, \ldots, x_n) \in \mathbb{R}^n$, $Q = (Q_1, \ldots, Q_m)$ and $G = (G_1, \ldots, G_m)$ are either real or complex valued functions, and ℓ, m and n are integers. The equation (1.1) is studied under certain boundary conditions, for example,

- periodic boundary conditions, e.g. Q is periodic in each component of x with period 2π,
- decay boundary conditions, e.g. $Q \to 0$ as $x \to \infty$.

Thus we have Cauchy problems for the equation (1.1), and we would like to pose the Cauchy problems in some Banach spaces $\mathcal{H}$, for example,

- $\mathcal{H}$ can be a Sobolev space H^k,
- $\mathcal{H}$ can be a Solobev space $H_{e,p}^k$ of even periodic functions.

We require that the problem is well-posed in $\mathcal{H}$, for example,

- for any $Q_0 \in \mathcal{H}$, there exists a unique solution $Q = Q(t, Q_0) \in C^0[(-\infty, \infty); \mathcal{H}]$ or $C^0[[0; \infty), \mathcal{H}]$ to the equation (1.1) such that $Q(0, Q_0) = Q_0$,
- for any fixed $t_0 \in (-\infty, \infty)$ or $[0, \infty)$, $Q(t_0, Q_0)$ is a C^r function of Q_0, for $Q_0 \in \mathcal{H}$ and some integer $r \geq 0$.

Example: Consider the integrable cubic nonlinear Schrödinger (NLS) equation,

$$(1.2) \qquad iq_t = q_{xx} + 2 \left[|\, q\, |^2 - \omega^2 \right] q \ ,$$

1

where $i = \sqrt{-1}$, $t \in \mathbb{R}^1$, $x \in \mathbb{R}^1$, q is a complex-valued function of (t, x), and ω is a real constant. We pose the periodic boundary condition,

$$q(t, x + 1) = q(t, x).$$

The Cauchy problem for equation (1.2) is posed in the Sobolev space H^1 of periodic functions,

$$\mathcal{H} \;\equiv\; \left\{ Q = (q, \bar{q}) \;\middle|\; q(x + 1) = q(x),\; q \in H^1_{[0,1]} : \text{the}\right.$$

$$\left. \text{Sobolev space } H^1 \text{ over the period interval } [0, 1] \right\},$$

and is well-posed [38] [32] [33].

Fact 1: For any $Q_0 \in \mathcal{H}$, there exists a unique solution $Q = Q(t, Q_0) \in C^0[(-\infty, \infty), \mathcal{H}]$ to the equation (1.2) such that $Q(0, Q_0) = Q_0$.

Fact 2: For any fixed $t_0 \in (-\infty, \infty)$, $Q(t_0, Q_0)$ is a C^2 function of Q_0, for $Q_0 \in \mathcal{H}$.

1.2. Phase Spaces and Flows

For finite dimensional dynamical systems, the phase spaces are often $\mathbb{R}^n$ or $\mathbb{C}^n$. For infinite dimensional dynamical systems, we take the Banach space $\mathcal{H}$ discussed in the previous section as the counterpart.

DEFINITION 1.1. We call the Banach space $\mathcal{H}$ in which the Cauchy problem for (1.1) is well-posed, a *phase space*. Define an operator F^t labeled by t as

$$Q(t, Q_0) = F^t(Q_0);$$

then $F^t : \mathcal{H} \to \mathcal{H}$ is called the *evolution operator* (or flow) for the system (1.1).

A point $p \in \mathcal{H}$ is called a *fixed point* if $F^t(p) = p$ for any t. Notice that here the fixed point p is in fact a function of x, which is the so-called stationary solution of (1.1). Let $q \in \mathcal{H}$ be a point; then $\ell_q \equiv \{F^t(q), \text{for all } t\}$ is called the orbit with initial point q. An orbit ℓ_q is called a *periodic orbit* if there exists a $T \in (-\infty, \infty)$ such that $F^T(q) = q$. An orbit ℓ_q is called a *homoclinic orbit* if there exists a point $q_* \in \mathcal{H}$ such that $F^t(q) \to q_*$, as $\mid t \mid \to \infty$, and q_* is called the asymptotic point of the homoclinic orbit. An orbit ℓ_q is called a *heteroclinic orbit* if there exist two different points $q_\pm \in \mathcal{H}$ such that $F^t(q) \to q_\pm$, as $t \to \pm\infty$, and $q_\pm$ are called the asymptotic points of the heteroclinic orbit. An orbit ℓ_q is said to be homoclinic to a submanifold W of $\mathcal{H}$ if $\inf_{Q \in W} \parallel F^t(q) - Q \parallel \to 0$, as $\mid t \mid \to \infty$.

Example 1: Consider the same Cauchy problem for the system (1.2). The fixed points of (1.2) satisfy the second order ordinary differential equation

$$(1.3) \qquad\qquad q_{xx} + 2\left[\mid q \mid^2 - \omega^2\right] q = 0.$$

In particular, there exists a circle of fixed points $q = \omega e^{i\gamma}$, where $\gamma \in [0, 2\pi]$. For simple periodic solutions, we have

$$(1.4) \qquad\qquad q = a e^{i\theta(t)}, \quad \theta(t) = -\left[2(a^2 - \omega^2)t - \gamma\right];$$

where $a > 0$, and $\gamma \in [0, 2\pi]$. For orbits homoclinic to the circles (1.4), we have

$$(1.5) \qquad q \;=\; \frac{1}{\Lambda}\left[\cos 2p - \sin p \, \operatorname{sech} \tau \cos 2\pi x - i \sin 2p \tanh \tau\right] a e^{i\theta(t)},$$

$$\Lambda = 1 + \sin p \, \operatorname{sech} \tau \cos 2\pi x,$$

where $\tau = 4\pi\sqrt{a^2 - \pi^2}\, t + \rho$, $p = \arctan\left[\frac{\sqrt{a^2-\pi^2}}{\pi}\right]$, $\rho \in (-\infty, \infty)$ is the Bäcklund parameter. Setting $a = \omega$ in (1.5), we have heteroclinic orbits asymptotic to points on the circle of fixed points. The expression (1.5) is generated from (1.4) through a Bäcklund-Darboux transformation [137].

Example 2: Consider the sine-Gordon equation,

$$u_{tt} - u_{xx} + \sin u = 0 \, ,$$

under the decay boundary condition that u belongs to the Schwartz class in x. The well-known "breather" solution,

$$(1.6) \qquad u(t,x) = 4\arctan\left[\frac{\tan\nu \cos[(\cos\nu)t]}{\cosh[(\sin\nu)x]}\right] ,$$

where ν is a parameter, is a periodic orbit. The expression (1.6) is generated from trivial solutions through a Bäcklund-Darboux transformation [59].

1.3. Invariant Submanifolds

Invariant submanifolds are the main objects in studying phase spaces. In phase spaces for partial differential equations, invariant submanifolds are often submanifolds with boundaries. Therefore, the following concepts on invariance are important.

DEFINITION 1.2 (Overflowing and Inflowing Invariance). A submanifold W with boundary ∂W is

- overflowing invariant if for any $t > 0$, $\bar{W} \subset F^t \circ W$, where $\bar{W} = W \cup \partial W$,
- inflowing invariant if any $t > 0$, $F^t \circ \bar{W} \subset W$,
- invariant if for any $t > 0$, $F^t \circ \bar{W} = \bar{W}$.

DEFINITION 1.3 (Local Invariance). A submanifold W with boundary ∂W is locally invariant if for any point $q \in W$, if $\bigcup\limits_{t\in[0,\infty)} F^t(q) \not\subset W$, then there exists $T \in (0, \infty)$ such that $\bigcup\limits_{t\in[0,T)} F^t(q) \subset W$, and $F^T(q) \in \partial W$; and if $\bigcup\limits_{t\in(-\infty,0]} F^t(q) \not\subset W$, then there exists $T \in (-\infty, 0)$ such that $\bigcup\limits_{t\in(T,0]} F^t(q) \subset W$, and $F^T(q) \in \partial W$.

Intuitively speaking, a submanifold with boundary is locally invariant if any orbit starting from a point inside the submanifold can only leave the submanifold through its boundary in both forward and backward time.

Example: Consider the linear equation,

$$(1.7) \qquad iq_t = (1+i)q_{xx} + iq \, ,$$

where $i = \sqrt{-1}$, $t \in \mathbb{R}^1$, $x \in \mathbb{R}^1$, and q is a complex-valued function of (t,x), under periodic boundary condition,

$$q(x+1) = q(x) \, .$$

Let $q = e^{\Omega_j t + i k_j x}$; then

$$\Omega_j = (1 - k_j^2) + i\, k_j^2 \, ,$$

where $k_j = 2j\pi$, $(j \in \mathbb{Z})$. $\Omega_0 = 1$, and when $|j| > 0$, $R_e\{\Omega_j\} < 0$. We take the H^1 space of periodic functions of period 1 to be the phase space. Then the submanifold

$$W_0 = \left\{ q \in H^1 \,\middle|\, q = c_0,\ c_0 \text{ is complex and } \| q \| < 1 \right\}$$

is an outflowing invariant submanifold, the submanifold

$$W_1 = \left\{ q \in H^1 \ \middle|\ q = c_1 e^{ik_1 x}, \quad c_1 \text{ is complex, and } \| q \| < 1 \right\}$$

is an inflowing invariant submanifold, and the submanifold

$$W = \left\{ q \in H^1 \ \middle|\ q = c_0 + c_1 e^{ik_1 x}, \quad c_0 \text{ and } c_1 \text{ are complex, and } \| q \| < 1 \right\}$$

is a locally invariant submanifold. The *unstable subspace* is given by

$$W^{(u)} = \left\{ q \in H^1 \ \middle|\ q = c_0, \quad c_0 \text{ is complex} \right\},$$

and the *stable subspace* is given by

$$W^{(s)} = \left\{ q \in H^1 \ \middle|\ q = \sum_{j \in Z/\{0\}} c_j \, e^{ik_j x}, \quad c_j\text{'s are complex} \right\}.$$

Actually, a good way to view the partial differential equation (1.7) as defining an infinite dimensional dynamical system is through Fourier transform, let

$$q(t,x) = \sum_{j \in Z} c_j(t) e^{ik_j x} \, ;$$

then $c_j(t)$ satisfy

$$\dot{c}_j = \left[(1 - k_j^2) + i k_j^2 \right] c_j, \quad j \in \mathbb{Z} \, ;$$

which is a system of infinitely many ordinary differential equations.

1.4. Poincaré Sections and Poincaré Maps

In the infinite dimensional phase space $\mathcal{H}$, Poincaré sections can be defined in a similar fashion as in a finite dimensional phase space. Let l_q be a periodic or homoclinic orbit in $\mathcal{H}$ under a flow F^t, and q_* be a point on l_q, then the Poincaré section Σ can be defined to be any codimension 1 subspace which has a transversal intersection with l_q at q_*. Then the flow F^t will induce a Poincaré map P in the neighborhood of q_* in Σ_0. Phase blocks, e.g. Smale horseshoes, can be defined using the norm.

CHAPTER 2

Soliton Equations as Integrable Hamiltonian PDEs

2.1. A Brief Summary

Soliton equations are integrable Hamiltonian partial differential equations. For example, the Korteweg-de Vries (KdV) equation

$$u_t = -6uu_x - u_{xxx} \; ,$$

where u is a real-valued function of two variables t and x, can be rewritten in the Hamiltonian form

$$u_t = \partial_x \frac{\delta H}{\delta u} \; ,$$

where

$$H = \int \left[\frac{1}{2} u_x^2 - u^3 \right] dx \; ,$$

under either periodic or decay boundary conditions. It is integrable in the classical Liouville sense, i.e., there exist enough functionally independent constants of motion. These constants of motion can be generated through isospectral theory or Bäcklund transformations [8]. The level sets of these constants of motion are elliptic tori [178] [154] [153] [68].

There exist soliton equations which possess level sets which are normally hyperbolic, for example, the focusing cubic nonlinear Schrödinger equation [137],

$$iq_t = q_{xx} + 2|q|^2 q \; ,$$

where $i = \sqrt{-1}$ and q is a complex-valued function of two variables t and x; the sine-Gordon equation [157],

$$u_{tt} = u_{xx} + \sin u \; ,$$

where u is a real-valued function of two variables t and x, etc.

Hyperbolic foliations are very important since they are the sources of chaos when the integrable systems are under perturbations. We will investigate the hyperbolic foliations of three typical types of soliton equations: (i). (1+1)-dimensional soliton equations represented by the focusing cubic nonlinear Schrödinger equation, (ii). soliton lattices represented by the focusing cubic nonlinear Schrödinger lattice, (iii). (1+2)-dimensional soliton equations represented by the Davey-Stewartson II equation.

REMARK 2.1. For those soliton equations which have only elliptic level sets, the corresponding representatives can be chosen to be the KdV equation for (1+1)-dimensional soliton equations, the Toda lattice for soliton lattices, and the KP equation for (1+2)-dimensional soliton equations.

Soliton equations are canonical equations which model a variety of physical phenomena, for example, nonlinear wave motions, nonlinear optics, plasmas, vortex

5

dynamics, etc. [5] [1]. Other typical examples of such integrable Hamiltonian partial differential equations are, e.g., the defocusing cubic nonlinear Schrödinger equation,

$$iq_t = q_{xx} - 2|q|^2 q \ ,$$

where $i = \sqrt{-1}$ and q is a complex-valued function of two variables t and x; the modified KdV equation,

$$u_t = \pm 6u^2 u_x - u_{xxx} \ ,$$

where u is a real-valued function of two variables t and x; the sinh-Gordon equation,

$$u_{tt} = u_{xx} + \sinh u \ ,$$

where u is a real-valued function of two variables t and x; the three-wave interaction equations,

$$\frac{\partial u_i}{\partial t} + a_i \frac{\partial u_i}{\partial x} = b_i \bar{u}_j \bar{u}_k \ ,$$

where $i, j, k = 1, 2, 3$ are cyclically permuted, a_i and b_i are real constants, u_i are complex-valued functions of t and x; the Boussinesq equation,

$$u_{tt} - u_{xx} + (u^2)_{xx} \pm u_{xxxx} = 0 \ ,$$

where u is a real-valued function of two variables t and x; the Toda lattice,

$$\partial^2 u_n / \partial t^2 = \exp\left\{-(u_n - u_{n-1})\right\} - \exp\left\{-(u_{n+1} - u_n)\right\} \ ,$$

where u_n's are real variables; the focusing cubic nonlinear Schrödinger lattice,

$$i\frac{\partial q_n}{\partial t} = (q_{n+1} - 2q_n + q_{n-1}) + |q_n|^2 (q_{n+1} + q_{n-1}) \ ,$$

where q_n's are complex variables; the Kadomtsev-Petviashvili (KP) equation,

$$(u_t + 6uu_x + u_{xxx})_x = \pm 3u_{yy} \ ,$$

where u is a real-valued function of three variables t, x and y; the Davey-Stewartson II equation,

$$\begin{cases} i\partial_t q = [\partial_x^2 - \partial_y^2]q + [2|q|^2 + u_y]q \ , \\[2mm] [\partial_x^2 + \partial_y^2]u = -4\partial_y |q|^2 \ , \end{cases}$$

where $i = \sqrt{-1}$, q is a complex-valued function of three variables t, x and y; and u is a real-valued function of three variables t, x and y. For more complete list of soliton equations, see e.g. [5] [1].

The cubic nonlinear Schrödinger equation is one of our main focuses in this book, which can be written in the Hamiltonian form,

$$iq_t = \frac{\delta H}{\delta \bar{q}} \ ,$$

where

$$H = \int [-|q_x|^2 \pm |q|^4] dx \ ,$$

under periodic boundary conditions. Its phase space is defined as

$$\mathcal{H}^k \equiv \left\{ \vec{q} = \begin{pmatrix} q \\ r \end{pmatrix} \ \middle| \ r = -\bar{q}, \ q(x+1) = q(x), \right.$$

$$\left. q \in H^k_{[0,1]} : \text{ the Sobolev space } H^k \text{ over the period interval } [0,1] \right\} \ .$$

REMARK 2.2. It is interesting to notice that the cubic nonlinear Schrödinger equation can also be written in Hamiltonian form in spatial variable, i.e.,

$$q_{xx} = iq_t \pm 2|q|^2 q \ ,$$

can be written in Hamiltonian form. Let $p = q_x$; then

$$\frac{\partial}{\partial x} \begin{pmatrix} q \\ \bar{q} \\ \bar{p} \\ p \end{pmatrix} = J \begin{pmatrix} \frac{\delta H}{\delta q} \\ \frac{\delta H}{\delta \bar{q}} \\ \frac{\delta H}{\delta \bar{p}} \\ \frac{\delta H}{\delta p} \end{pmatrix} \ ,$$

where

$$J = \begin{pmatrix} 0 & 0 & 1 & 0 \\ 0 & 0 & 0 & 1 \\ -1 & 0 & 0 & 0 \\ 0 & -1 & 0 & 0 \end{pmatrix} \ ,$$

$$H = \int [|p|^2 \mp |q|^4 - \frac{i}{2}(q_t \bar{q} - \bar{q}_t q)]dt \ ,$$

under decay or periodic boundary conditions. We do not know whether or not other soliton equations have this property.

2.2. A Physical Application of the Nonlinear Schrödinger Equation

The cubic nonlinear Schrödinger (NLS) equation has many different applications, i.e. it describes many different physical phenomena, and that is why it is called a canonical equation. Here, as an example, we show how the NLS equation describes the motion of a vortex filament – the beautiful Hasimoto derivation [82]. Vortex filaments in an inviscid fluid are known to preserve their identities. The motion of a very thin isolated vortex filament $\vec{X} = \vec{X}(s,t)$ of radius ϵ in an incompressible inviscid unbounded fluid by its own induction is described asymptotically by

$$(2.1) \qquad \partial \vec{X}/\partial t = G\kappa \vec{b} \ ,$$

where s is the length measured along the filament, t is the time, κ is the curvature, $\vec{b}$ is the unit vector in the direction of the binormal and G is the coefficient of local induction,

$$G = \frac{\Gamma}{4\pi}[\ln(1/\epsilon) + O(1)] \ ,$$

which is proportional to the circulation Γ of the filament and may be regarded as a constant if we neglect the second order term. Then a suitable choice of the units of time and length reduces (2.1) to the nondimensional form,

$$(2.2) \qquad \partial \vec{X}/\partial t = \kappa \vec{b} \ .$$

Equation (2.2) should be supplemented by the equations of differential geometry (the Frenet-Seret formulae)

$$(2.3) \qquad \partial \vec{X}/\partial s = \vec{t} \ , \quad \partial \vec{t}/\partial s = \kappa \vec{n} \ , \quad \partial \vec{n}/\partial s = \tau \vec{b} - \kappa \vec{t} \ , \quad \partial \vec{b}/\partial s = -\tau \vec{n} \ ,$$

where τ is the torsion and $\vec{t}$, $\vec{n}$ and $\vec{b}$ are the tangent, the principal normal and the binormal unit vectors. The last two equations imply that

$$(2.4) \qquad \partial(\vec{n} + i\vec{b})/\partial s = -i\tau(\vec{n} + i\vec{b}) - \kappa\vec{t} ,$$

which suggests the introduction of new variables

$$(2.5) \qquad \vec{N} = (\vec{n} + i\vec{b})\exp\left\{i\int_0^s \tau ds\right\} ,$$

and

$$(2.6) \qquad q = \kappa\exp\left\{i\int_0^s \tau ds\right\} .$$

Then from (2.3) and (2.4), we have

$$(2.7) \qquad \partial\vec{N}/\partial s = -q\vec{t} , \quad \partial\vec{t}/\partial s = \text{Re}\{q\overline{\vec{N}}\} = \frac{1}{2}(\bar{q}\vec{N} + q\overline{\vec{N}}) .$$

We are going to use the relation $\frac{\partial^2 \vec{N}}{\partial s \partial t} = \frac{\partial^2 \vec{N}}{\partial t \partial s}$ to derive an equation for q. For this we need to know $\partial\vec{t}/\partial t$ and $\partial\vec{N}/\partial t$ besides equations (2.7). From (2.2) and (2.3), we have

$$\partial\vec{t}/\partial t = \frac{\partial^2 \vec{X}}{\partial s \partial t} = \partial(\kappa\vec{b})/\partial s = (\partial\kappa/\partial s)\vec{b} - \kappa\tau\vec{n}$$
$$= \kappa\,\text{Re}\{(\frac{1}{\kappa}\partial\kappa/\partial s + i\tau)(\vec{b} + i\vec{n})\} ,$$

i.e.

$$(2.8) \qquad \partial\vec{t}/\partial t = \text{Re}\{i(\partial q/\partial s)\overline{\vec{N}}\} = \frac{1}{2}i[(\partial q/\partial s)\overline{\vec{N}} - (\partial q/\partial s)^-\vec{N}] .$$

We can write the equation for $\partial\vec{N}/\partial t$ in the following form:

$$(2.9) \qquad \partial\vec{N}/\partial t = \alpha\vec{N} + \beta\overline{\vec{N}} + \gamma\vec{t} ,$$

where α, β and γ are complex coefficients to be determined.

$$\alpha + \bar{\alpha} = \frac{1}{2}[\partial\vec{N}/\partial t \cdot \overline{\vec{N}} + \partial\overline{\vec{N}}/\partial t \cdot \vec{N}]$$
$$= \frac{1}{2}\partial(\vec{N} \cdot \overline{\vec{N}})/\partial t = 0 ,$$

i.e. $\alpha = iR$ where R is an unknown real function.

$$\beta = \frac{1}{2}\partial\vec{N}/\partial t \cdot \vec{N} = \frac{1}{4}\partial(\vec{N} \cdot \vec{N})/\partial t = 0 ,$$
$$\gamma = -\vec{N} \cdot \partial\vec{t}/\partial t = -i\partial q/\partial s .$$

Thus

$$(2.10) \qquad \partial\vec{N}/\partial t = i[R\vec{N} - (\partial q/\partial s)\vec{t}] .$$

From (2.7), (2.10) and (2.8), we have

$$\frac{\partial^2 \vec{N}}{\partial s \partial t} = -(\partial q/\partial t)\vec{t} - q\partial \vec{t}/\partial t$$

$$= -(\partial q/\partial t)\vec{t} - \frac{1}{2}iq[(\partial q/\partial s)\overline{\vec{N}} - (\partial q/\partial s)^{-}\vec{N}] ,$$

$$\frac{\partial^2 \vec{N}}{\partial t \partial s} = i[(\partial R/\partial s)\vec{N} - Rq\vec{t} - (\partial^2 q/\partial s^2)\vec{t}$$

$$- \frac{1}{2}(\partial q/\partial s)(\bar{q}\vec{N} + q\overline{\vec{N}})] .$$

Thus, we have

(2.11)
$$\partial q/\partial t = i[\partial^2 q/\partial s^2 + Rq] ,$$

and

(2.12)
$$\frac{1}{2}q\partial \bar{q}/\partial s = \partial R/\partial s - \frac{1}{2}(\partial q/\partial s)\bar{q} .$$

The comparison of expressions for $\frac{\partial^2 \vec{t}}{\partial s \partial t}$ from (2.7) and (2.8) leads only to (2.11). Solving (2.12), we have

(2.13)
$$R = \frac{1}{2}(|q|^2 + A) ,$$

where A is a real-valued function of t only. Thus we have the cubic nonlinear Schrödinger equation for q:

$$-i\partial q/\partial t = \partial^2 q/\partial s^2 + \frac{1}{2}(|q|^2 + A)q .$$

The term Aq can be transformed away by defining the new variable

$$\tilde{q} = q\exp[-\frac{1}{2}i\int_0^t A(t)dt] .$$

CHAPTER 3

Figure-Eight Structures

For finite-dimensional Hamiltonian systems, figure-eight structures are often given by singular level sets. These singular level sets are also called separatrices. Expressions for such figure-eight structures can be obtained by setting the Hamiltonian and/or other constants of motion to special values. For partial differential equations, such an approach is not feasible. For soliton equations, expressions for figure-eight structures can be obtained via Bäcklund-Darboux transformations [137] [118] [121].

3.1. 1D Cubic Nonlinear Schrödinger (NLS) Equation

We take the focusing nonlinear Schrödinger equation (NLS) as our first example to show how to construct figure-eight structures. If one starts from the conservational laws of the NLS, it turns out that it is very elusive to get the separatrices. On the contrary, starting from the Bäcklund-Darboux transformation to be presented, one can find the separatrices rather easily. We consider the NLS

$$(3.1) \qquad\qquad iq_t = q_{xx} + 2|q|^2 q ,$$

under periodic boundary condition $q(x + 2\pi) = q(x)$. The NLS is an integrable system by virtue of the Lax pair [212],

$$(3.2) \qquad\qquad \varphi_x \;=\; U\varphi ,$$
$$(3.3) \qquad\qquad \varphi_t \;=\; V\varphi ,$$

where

$$U = i\lambda\sigma_3 + i \begin{pmatrix} 0 & q \\ -r & 0 \end{pmatrix} ,$$

$$V = 2i\lambda^2\sigma_3 + iqr\sigma_3 + \begin{pmatrix} 0 & 2i\lambda q + q_x \\ -2i\lambda r + r_x & 0 \end{pmatrix} ,$$

where σ_3 denotes the third Pauli matrix $\sigma_3 = \text{diag}(1, -1)$, $r = -\bar{q}$, and λ is the spectral parameter. If q satisfies the NLS, then the compatibility of the over determined system (3.2, 3.3) is guaranteed. Let $M = M(x)$ be the fundamental matrix solution to the ODE (3.2), $M(0)$ is the 2×2 identity matrix. We introduce the so-called transfer matrix $T = T(\lambda, \vec{q})$ where $\vec{q} = (q, -\bar{q})$, $T = M(2\pi)$.

LEMMA 3.1. *Let $Y(x)$ be any solution to the ODE (3.2), then*

$$Y(2n\pi) = T^n \, Y(0) .$$

Proof: Since $M(x)$ is the fundamental matrix,

$$Y(x) = M(x) \, Y(0) .$$

11

Thus,

$$Y(2\pi) = T\, Y(0)\,.$$

Assume that

$$Y(2l\pi) = T^l\, Y(0)\,.$$

Notice that $Y(x + 2l\pi)$ also solves the ODE (3.2); then

$$Y(x + 2l\pi) = M(x)\, Y(2l\pi)\,;$$

thus,

$$Y(2(l+1)\pi) = T\, Y(2l\pi) = T^{l+1}\, Y(0)\,.$$

The lemma is proved. Q.E.D.

DEFINITION 3.2. We define the Floquet discriminant Δ as,

$$\Delta(\lambda, \vec{q}) = \ \text{trace}\ \{T(\lambda, \vec{q})\}\,.$$

We define the periodic and anti-periodic points $\lambda^{(p)}$ by the condition

$$|\Delta(\lambda^{(p)}, \vec{q})| = 2\,.$$

We define the critical points $\lambda^{(c)}$ by the condition

$$\left.\frac{\partial \Delta(\lambda, \vec{q})}{\partial \lambda}\right|_{\lambda = \lambda^{(c)}} = 0\,.$$

A multiple point, denoted $\lambda^{(m)}$, is a critical point for which

$$|\Delta(\lambda^{(m)}, \vec{q})| = 2.$$

The algebraic multiplicity of $\lambda^{(m)}$ is defined as the order of the zero of $\Delta(\lambda) \pm 2$. Ususally it is 2, but it can exceed 2; when it does equal 2, we call the multiple point a double point, and denote it by $\lambda^{(d)}$. The geometric multiplicity of $\lambda^{(m)}$ is defined as the maximum number of linearly independent solutions to the ODE (3.2), and is either 1 or 2.

Let $q(x, t)$ be a solution to the NLS (3.1) for which the linear system (3.2) has a complex double point ν of geometric multiplicity 2. We denote two linearly independent solutions of the Lax pair (3.2,3.3) at $\lambda = \nu$ by (ϕ^+, ϕ^-). Thus, a general solution of the linear systems at (q, ν) is given by

$$(3.4) \qquad \phi(x, t) = c_+\phi^+ \ + \ c_-\phi^-\,.$$

We use ϕ to define a Gauge matrix [**180**] G by

$$(3.5) \qquad G = G(\lambda; \nu; \phi) = N \begin{pmatrix} \lambda - \nu & 0 \\ 0 & \lambda - \bar{\nu} \end{pmatrix} N^{-1}\,,$$

where

$$(3.6) \qquad N = \begin{pmatrix} \phi_1 & -\bar{\phi}_2 \\ \phi_2 & \bar{\phi}_1 \end{pmatrix}\,.$$

Then we define Q and Ψ by

$$(3.7) \qquad Q(x, t) = q(x, t) \ + \ 2(\nu - \bar{\nu})\,\frac{\phi_1 \bar{\phi}_2}{\phi_1 \bar{\phi}_1 + \phi_2 \bar{\phi}_2}$$

and

$$(3.8) \qquad \Psi(x, t; \lambda) \ = \ G(\lambda; \nu; \phi)\, \psi(x, t; \lambda)$$

where ψ solves the Lax pair (3.2,3.3) at (q,ν). Formulas (3.7) and (3.8) are the Bäcklund-Darboux transformations for the potential and eigenfunctions, respectively. We have the following [180] [137],

THEOREM 3.3. *Let $q(x,t)$ be a solution to the NLS equation (3.1), for which the linear system (3.2) has a complex double point ν of geometric multiplicity 2, with eigenbasis (ϕ^+, ϕ^-) for the Lax pair (3.2,3.3), and define $Q(x,t)$ and $\Psi(x,t;\lambda)$ by (3.7) and (3.8). Then*

(1) *$Q(x,t)$ is an solution of NLS, with spatial period 2π,*
(2) *Q and q have the same Floquet spectrum,*
(3) *$Q(x,t)$ is homoclinic to $q(x,t)$ in the sense that $Q(x,t) \longrightarrow q_{\theta_\pm}(x,t)$, expontentially as $\exp(-\sigma_\nu|t|)$, as $t \longrightarrow \pm\infty$, where $q_{\theta_\pm}$ is a "torus translate" of q, σ_ν is the nonvanishing growth rate associated to the complex double point ν, and explicit formulas exist for this growth rate and for the translation parameters $\theta_\pm$,*
(4) *$\Psi(x,t;\lambda)$ solves the Lax pair (3.2,3.3) at (Q,λ).*

This theorem is quite general, constructing homoclinic solutions from a wide class of starting solutions $q(x,t)$. It's proof is one of direct verification [118].

We emphasize several qualitative features of these homoclinic orbits: (i) $Q(x,t)$ is homoclinic to a torus which itself possesses rather complicated spatial and temporal structure, and is not just a fixed point. (ii) Nevertheless, the homoclinic orbit typically has still more complicated spatial structure than its "target torus". (iii) When there are several complex double points, each with nonvanishing growth rate, one can iterate the Bäcklund-Darboux transformations to generate more complicated homoclinic orbits. (iv) The number of complex double points with nonvanishing growth rates counts the dimension of the unstable manifold of the critical torus in that two unstable directions are coordinatized by the complex ratio c_+/c_-. Under even symmetry only one real dimension satisfies the constraint of evenness, as will be clearly illustrated in the following example. (v) These Bäcklund-Darboux formulas provide global expressions for the stable and unstable manifolds of the critical tori, which represent figure-eight structures.

Example: As a concrete example, we take $q(x,t)$ to be the special solution

$$(3.9) \qquad q_c = c\exp\left\{-i[2c^2t + \gamma]\right\} .$$

Solutions of the Lax pair (3.2,3.3) can be computed explicitly:

$$(3.10) \qquad \phi^{(\pm)}(x,t;\lambda) = e^{\pm i\kappa(x+2\lambda t)} \begin{pmatrix} ce^{-i(2c^2t+\gamma)/2} \\ (\pm\kappa - \lambda)e^{i(2c^2t+\gamma)/2} \end{pmatrix} ,$$

where

$$\kappa = \kappa(\lambda) = \sqrt{c^2 + \lambda^2} .$$

With these solutions one can construct the fundamental matrix

$$(3.11) \qquad M(x;\lambda;q_c) = \begin{bmatrix} \cos\kappa x + i\frac{\lambda}{\kappa}\sin\kappa x & i\frac{q_c}{\kappa}\sin\kappa x \\ i\frac{\overline{q_c}}{\kappa}\sin\kappa x & \cos\kappa x - i\frac{\lambda}{\kappa}\sin\kappa x \end{bmatrix} ,$$

from which the Floquet discriminant can be computed:

$$(3.12) \qquad \Delta(\lambda;q_c) = 2\cos(2\kappa\pi) .$$

From Δ, spectral quantities can be computed:

(1) simple periodic points: $\lambda^\pm = \pm i\, c$,

(2) double points: $\kappa(\lambda_j^{(d)}) = j/2$, $j \in \mathbb{Z}$, $j \neq 0$,

(3) critical points: $\lambda_j^{(c)} = \lambda_j^{(d)}$, $j \in \mathbb{Z}$, $j \neq 0$,

(4) simple periodic points: $\lambda_0^{(c)} = 0$.

For this spectral data, there are 2N purely imaginary double points,

$$(3.13) \qquad (\lambda_j^{(d)})^2 = j^2/4 - c^2, \quad j = 1, 2, \cdots, N;$$

where

$$\left[N^2/4 - c^2 \right] < 0 < \left[(N+1)^2/4 - c^2 \right] .$$

From this spectral data, the homoclinic orbits can be explicitly computed through Bäcklund-Darboux transformation. Notice that to have temporal growth (and decay) in the eigenfunctions (3.10), one needs λ to be complex. Notice also that the Bäcklund-Darboux transformation is built with quadratic products in ϕ, thus choosing $\nu = \lambda_j^{(d)}$ will guarantee periodicity of Q in x. When $N = 1$, the Bäcklund-Darboux transformation at one purely imaginary double point $\lambda_1^{(d)}$ yields $Q = Q(x, t; c, \gamma; c_+/c_-)$ [**137**]:

$$(3.14) \qquad \begin{aligned} Q &= \left[\cos 2p - \sin p \ \operatorname{sech}\tau \ \cos(x + \vartheta) - i \sin 2p \tanh \tau \right] \\ &\quad \left[1 + \sin p \ \operatorname{sech}\tau \ \cos(x + \vartheta) \right]^{-1} c e^{-i(2c^2 t + \gamma)} \\ &\to e^{\mp 2ip} c e^{-i(2c^2 t + \gamma)} \quad \text{as} \ \rho \to \mp\infty, \end{aligned}$$

where $c_+/c_- \equiv \exp(\rho + i\beta)$ and p is defined by $1/2 + i\sqrt{c^2 - 1/4} = c\exp(ip)$, $\tau \equiv \sigma t - \rho$, and $\vartheta = p - (\beta + \pi/2)$.

Several points about this homoclinic orbit need to be made:

(1) The orbit depends only upon the ratio c_+/c_-, and not upon c_+ and c_- individually.

(2) Q is homoclinic to the plane wave orbit; however, a phase shift of $-4p$ occurs when one compares the asymptotic behavior of the orbit as $t \to -\infty$ with its behavior as $t \to +\infty$.

(3) For small p, the formula for Q becomes more transparent:

$$Q \simeq \left[(\cos 2p - i \sin 2p \tanh \tau) - 2 \sin p \ \operatorname{sech} \tau \cos(x + \vartheta) \right] c e^{-i(2c^2 t + \gamma)}.$$

(4) An evenness constraint on Q in x can be enforced by restricting the phase ϕ to be one of two values

$$\phi = 0, \pi. \qquad\qquad \text{(evenness)}$$

In this manner, the even symmetry disconnects the level set. Each component constitutes one loop of the figure eight. While the target q is independent of x, each of these loops has x dependence through the $\cos(x)$. One loop has exactly this dependence and can be interpreted as a spatial excitation located near $x = 0$, while the second loop has the dependence $\cos(x - \pi)$, which we interpret as spatial structure located near $x = \pi$. In this example, the disconnected nature of the level set is clearly related to distinct spatial structures on the individual loops. See Figure 1 for an illustration.

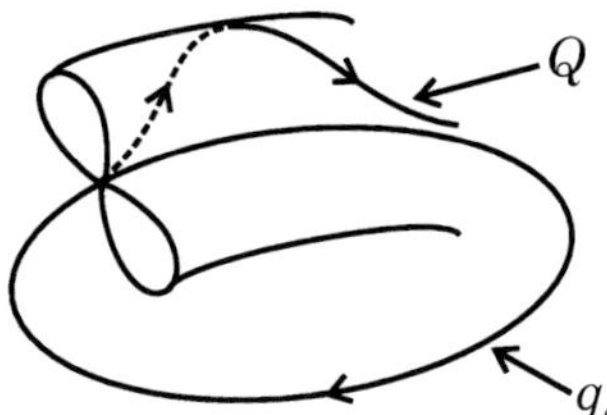

FIGURE 1. An illustration of the figure-eight structure.

(5) Direct calculation shows that the transformation matrix $M(1; \lambda_1^{(d)}; Q)$ is similar to a Jordan form when $t \in (-\infty, \infty)$,

$$M(1; \lambda_1^{(d)}; Q) \sim \begin{pmatrix} -1 & 1 \\ 0 & -1 \end{pmatrix} ,$$

and when $t \to \pm\infty$, $M(1; \lambda_1^{(d)}; Q) \longrightarrow -I$ (the negative of the 2x2 identity matrix). Thus, when t is finite, the algebraic multiplicity $(= 2)$ of $\lambda = \lambda_1^{(d)}$ with the potential Q is greater than the geometric multiplicity $(= 1)$.

In this example the dimension of the loops need not be one, but is determined by the number of purely imaginary double points which in turn is controlled by the amplitude c of the plane wave target and by the spatial period. (The dimension of the loops increases linearly with the spatial period.) When there are several complex double points, Bäcklund-Darboux transformations must be iterated to produce complete representations. Thus, Bäcklund-Darboux transformations give global representations of the figure-eight structures.

3.1.1. Linear Instability. The above figure-eight structure corresponds to the following linear instability of Benjamin-Feir type. Consider the uniform solution to the NLS (3.1),

$$q_c = ce^{i\theta(t)} , \qquad \theta(t) = -[2c^2t + \gamma] .$$

Let

$$q = [c + \tilde{q}]e^{i\theta(t)} ,$$

and linearize equation (3.1) at q_c, we have

$$i\tilde{q}_t = \tilde{q}_{xx} + 2c^2[\tilde{q} + \bar{\tilde{q}}] .$$

Assume that $\tilde{q}$ takes the form,

$$\tilde{q} = \left[A_j e^{\Omega_j t} + B_j e^{\bar{\Omega}_j t}\right] \cos k_j x ,$$

where $k_j = 2j\pi$, $(j = 0, 1, 2, \cdots)$, A_j and B_j are complex constants. Then,

$$\Omega_j^{(\pm)} = \pm k_j \sqrt{4c^2 - k_j^2} .$$

Thus, we have instabilities when $c > 1/2$.

3.1.2. Quadratic Products of Eigenfunctions. Quadratic products of eigenfunctions play a crucial role in characterizing the hyperbolic structures of soliton equations. Its importance lies in the following aspects: (i). Certain quadratic products of eigenfunctions solve the linearized soliton equation. (ii). Thus, they are the perfect candidates for building a basis to the invariant linear subbundles. (iii). Also, they signify the instability of the soliton equation. (iv). Most importantly, quadratic products of eigenfunctions can serve as Melnikov vectors, e.g., for Davey-Stewartson equation [**121**].

Consider the linearized NLS equation at any solution $q(t,x)$ written in the vector form:

$$i\partial_t(\delta q) \;=\; (\delta q)_{xx} + 2[q^2\overline{\delta q} + 2|q|^2\delta q] \;,$$

(3.15)

$$i\partial_t(\overline{\delta q}) \;=\; -(\overline{\delta q})_{xx} - 2[\bar{q}^2\delta q + 2|q|^2\overline{\delta q}] \;,$$

we have the following lemma [**137**].

LEMMA 3.4. *Let $\varphi^{(j)} = \varphi^{(j)}(t,x;\lambda,q)$ $(j=1,2)$ be any two eigenfunctions solving the Lax pair (3.2,3.3) at an arbitrary λ. Then*

$$\begin{pmatrix} \delta q \\ \overline{\delta q} \end{pmatrix}\;,\quad \begin{pmatrix} \varphi_1^{(1)}\varphi_1^{(2)} \\ \varphi_2^{(1)}\varphi_2^{(2)} \end{pmatrix}\;,\quad and \;\; S\begin{pmatrix} \varphi_1^{(1)}\varphi_1^{(2)} \\ \varphi_2^{(1)}\varphi_2^{(2)} \end{pmatrix}^{-}\;,\quad where\;\; S = \begin{pmatrix} 0 & 1 \\ 1 & 0 \end{pmatrix}$$

solve the same equation (3.15); thus

$$\Phi = \begin{pmatrix} \varphi_1^{(1)}\varphi_1^{(2)} \\ \varphi_2^{(1)}\varphi_2^{(2)} \end{pmatrix} + S\begin{pmatrix} \varphi_1^{(1)}\varphi_1^{(2)} \\ \varphi_2^{(1)}\varphi_2^{(2)} \end{pmatrix}^{-}$$

solves the equation (3.15) and satisfies the reality condition $\Phi_2 = \bar{\Phi}_1$.

Proof: Direct calculation leads to the conclusion. Q.E.D.

The periodicity condition $\Phi(x+2\pi) = \Phi(x)$ can be easily accomplished. For example, we can take $\varphi^{(j)}$ $(j=1,2)$ to be two linearly independent Bloch functions $\varphi^{(j)} = e^{\sigma_j x}\psi^{(j)}$ $(j=1,2)$, where $\sigma_2 = -\sigma_1$ and $\psi^{(j)}$ are periodic functions $\psi^{(j)}(x+2\pi) = \psi^{(j)}(x)$. Often we choose λ to be a double point of geometric multiplicity 2, so that $\varphi^{(j)}$ are already periodic or antiperiodic functions.

3.2. Discrete Cubic Nonlinear Schrödinger Equation

Consider the discrete focusing cubic nonlinear Schrödinger equation (DNLS)

$$(3.16)\qquad i\dot{q}_n = \frac{1}{h^2}[q_{n+1} - 2q_n + q_{n-1}] + |q_n|^2(q_{n+1}+q_{n-1}) - 2\omega^2 q_n \;,$$

under periodic and even boundary conditions,

$$q_{n+N} = q_n \;,\qquad q_{-n} = q_n \;,$$

where $i = \sqrt{-1}$, q_n's are complex variables, $n \in \mathbb{Z}$, ω is a positive parameter, $h = 1/N$, and N is a positive integer $N \geq 3$. The DNLS is integrable by virtue of the Lax pair [**2**]:

$$(3.17)\qquad\qquad\qquad \varphi_{n+1} = L_n^{(z)}\varphi_n \;,$$

$$(3.18)\qquad\qquad\qquad \dot{\varphi}_n = B_n^{(z)}\varphi_n \;,$$

where

$$L_n^{(z)} = \begin{pmatrix} z & ihq_n \\ ih\bar{q}_n & 1/z \end{pmatrix},$$

$$B_n^{(z)} = \frac{i}{h^2} \begin{pmatrix} b_n^{(1)} & -izhq_n + (1/z)ihq_{n-1} \\ -izh\bar{q}_{n-1} + (1/z)ih\bar{q}_n & b_n^{(4)} \end{pmatrix},$$

$$b_n^{(1)} = 1 - z^2 + 2i\lambda h - h^2 q_n \bar{q}_{n-1} + \omega^2 h^2,$$

$$b_n^{(4)} = 1/z^2 - 1 + 2i\lambda h + h^2 \bar{q}_n q_{n-1} - \omega^2 h^2,$$

and where $z = \exp(i\lambda h)$. Compatibility of the over-determined system (3.17,3.18) gives the "Lax representation"

$$\dot{L}_n = B_{n+1} L_n - L_n B_n$$

of the DNLS (3.16). Let $M(n)$ be the fundamental matrix solution to (3.17), the *Floquet discriminant* is defined as

$$\Delta = \text{trace } \{M(N)\} .$$

Let ψ^+ and ψ^- be any two solutions to (3.17), and let $W_n(\psi^+, \psi^-)$ be the Wronskian

$$W_n(\psi^+, \psi^-) = \psi_n^{(+,1)} \psi_n^{(-,2)} - \psi_n^{(+,2)} \psi_n^{(-,1)} .$$

One has

$$W_{n+1}(\psi^+, \psi^-) = \rho_n W_n(\psi^+, \psi^-) ,$$

where $\rho_n = 1 + h^2 |q_n|^2$, and

$$W_N(\psi^+, \psi^-) = D^2 \, W_0(\psi^+, \psi^-) ,$$

where $D^2 = \prod_{n=0}^{N-1} \rho_n$. Periodic and antiperiodic points $z^{(p)}$ are defined by

$$\Delta(z^{(p)}) = \pm 2D .$$

A critical point $z^{(c)}$ is defined by the condition

$$\left. \frac{d\Delta}{dz} \right|_{z=z^{(c)}} = 0.$$

A multiple point $z^{(m)}$ is a critical point which is also a periodic or antiperiodic point. The *algebraic multiplicity* of $z^{(m)}$ is defined as the order of the zero of $\Delta(z) \pm 2D$. Usually it is 2, but it can exceed 2; when it does equal 2, we call the multiple point a *double point,* and denote it by $z^{(d)}$. The *geometric multiplicity* of $z^{(m)}$ is defined as the dimension of the periodic (or antiperiodic) eigenspace of (3.17) at $z^{(m)}$, and is either 1 or 2.

Fix a solution $q_n(t)$ of the DNLS (3.16), for which (3.17) has a double point $z^{(d)}$ of geometric multiplicity 2, which is not on the unit circle. We denote two linearly independent solutions (Bloch functions) of the discrete Lax pair (3.17,3.18) at $z = z^{(d)}$ by (ϕ_n^+, ϕ_n^-). Thus, a general solution of the discrete Lax pair (3.17;3.18) at $(q_n(t), z^{(d)})$ is given by

$$\phi_n = c^+ \phi_n^+ + c^- \phi_n^-,$$

where c^+ and c^- are complex parameters. We use ϕ_n to define a transformation matrix Γ_n by

$$\Gamma_n = \begin{pmatrix} z + (1/z)a_n & b_n \\ c_n & -1/z + zd_n \end{pmatrix},$$

where,

$$a_n = \frac{z^{(d)}}{(\bar{z}^{(d)})^2 \Delta_n}\left[|\phi_{n2}|^2 + |z^{(d)}|^2|\phi_{n1}|^2\right],$$

$$d_n = -\frac{1}{z^{(d)}\Delta_n}\left[|\phi_{n2}|^2 + |z^{(d)}|^2|\phi_{n1}|^2\right],$$

$$b_n = \frac{|z^{(d)}|^4 - 1}{(\bar{z}^{(d)})^2 \Delta_n}\phi_{n1}\bar{\phi}_{n2},$$

$$c_n = \frac{|z^{(d)}|^4 - 1}{z^{(d)}\bar{z}^{(d)}\Delta_n}\bar{\phi}_{n1}\phi_{n2},$$

$$\Delta_n = -\frac{1}{\bar{z}^{(d)}}\left[|\phi_{n1}|^2 + |z^{(d)}|^2|\phi_{n2}|^2\right].$$

From these formulae, we see that

$$\bar{a}_n = -d_n, \quad \bar{b}_n = c_n.$$

Then we define Q_n and Ψ_n by

$$(3.19) \qquad\qquad Q_n \equiv \frac{i}{h}b_{n+1} - a_{n+1}q_n$$

and

$$(3.20) \qquad\qquad \Psi_n(t;z) \equiv \Gamma_n(z;z^{(d)};\phi_n)\psi_n(t;z)$$

where ψ_n solves the discrete Lax pair (3.17,3.18) at $(q_n(t),z)$. Formulas (3.19) and (3.20) are the Bäcklund-Darboux transformations for the potential and eigenfunctions, respectively. We have the following theorem [118].

THEOREM 3.5. *Let $q_n(t)$ denote a solution of the DNLS (3.16), for which (3.17) has a double point $z^{(d)}$ of geometric multiplicity 2, which is not on the unit circle. We denote two linearly independent solutions of the discrete Lax pair (3.17,3.18) at $(q_n, z^{(d)})$ by (ϕ_n^+, ϕ_n^-). We define $Q_n(t)$ and $\Psi_n(t;z)$ by (3.19) and (3.20). Then*

(1) *$Q_n(t)$ is also a solution of the DNLS (3.16). (The eveness of Q_n can be obtained by choosing the complex Bäcklund parameter c^+/c^- to lie on a certain curve, as shown in the example below.)*

(2) *$\Psi_n(t;z)$ solves the discrete Lax pair (3.17,3.18) at $(Q_n(t),z)$.*

(3) *$\Delta(z;Q_n) = \Delta(z;q_n)$, for all $z \in C$.*

(4) *$Q_n(t)$ is homoclinic to $q_n(t)$ in the sense that $Q_n(t) \to e^{i\theta_\pm}q_n(t)$, exponentially as $\exp(-\sigma|t|)$ as $t \to \pm\infty$. Here $\theta_\pm$ are the phase shifts, σ is a nonvanishing growth rate associated to the double point $z^{(d)}$, and explicit formulas can be developed for this growth rate and for the phase shifts $\theta_\pm$.*

Example: We start with the uniform solution of (3.16)

$$(3.21) \qquad q_n = q_c \, , \ \forall n; \quad q_c = a\exp\left\{-i[2(a^2 - \omega^2)t - \gamma]\right\}.$$

We choose the amplitude a in the range

$$N\tan\frac{\pi}{N} < a < N\tan\frac{2\pi}{N} \, , \quad \text{when } N > 3 \, ,$$

$$(3.22)$$

$$3\tan\frac{\pi}{3} < a < \infty \, , \quad \text{when } N = 3 \, ;$$

so that there is only one set of quadruplets of double points which are not on the unit circle, and denote one of them by $z = z_1^{(d)} = z_1^{(c)}$ which corresponds to $\beta = \pi/N$. The homoclinic orbit Q_n is given by

$$(3.23) \qquad Q_n = q_c(\hat{E}_{n+1})^{-1}\left[\hat{A}_{n+1} - 2\cos\beta\sqrt{\rho\cos^2\beta - 1}\,\hat{B}_{n+1}\right],$$

where

$$\hat{E}_n = ha\cos\beta + \sqrt{\rho\cos^2\beta - 1}\ \mathrm{sech}\,[2\mu t + 2p]\cos[(2n-1)\beta + \vartheta]\ ,$$

$$\hat{A}_{n+1} = ha\cos\beta + \sqrt{\rho\cos^2\beta - 1}\ \mathrm{sech}\,[2\mu t + 2p]\cos[(2n+3)\beta + \vartheta]\ ,$$

$$\hat{B}_{n+1} = \cos\varphi + i\sin\varphi\tanh[2\mu t + 2p] + \ \mathrm{sech}\,[2\mu t + 2p]\cos[2(n+1)\beta + \vartheta]\ ,$$

$$\beta = \pi/N\ ,\quad \rho = 1 + h^2 a^2\ ,\quad \mu = 2h^{-2}\sqrt{\rho}\sin\beta\sqrt{\rho\cos^2\beta - 1}\ ,$$

$$h = 1/N,\quad c_+/c_- = ie^{2p}e^{i\vartheta}\ ,\quad \vartheta \in [0, 2\pi]\ ,\quad p \in (-\infty, \infty)\ ,$$

$$z_1^{(d)} = \sqrt{\rho}\cos\beta + \sqrt{\rho\cos^2\beta - 1}\ ,\quad \theta(t) = (a^2 - \omega^2)t - \gamma/2\ ,$$

$$\sqrt{\rho\cos^2\beta - 1} + i\sqrt{\rho}\sin\beta = hae^{i\varphi}\ ,$$

where $\varphi = \sin^{-1}[\sqrt{\rho}(ha)^{-1}\sin\beta]$, $\varphi \in (0, \pi/2)$.

Next we study the "evenness" condition: $Q_{-n} = Q_n$. It turns out that the choices $\vartheta = -\beta$, $-\beta + \pi$ in the formula of Q_n lead to the evenness of Q_n in n. In terms of figure eight structure of Q_n, $\vartheta = -\beta$ corresponds to one ear of the figure eight, and $\vartheta = -\beta + \pi$ corresponds to the other ear. The even formula for Q_n is given by,

$$(3.24) \qquad Q_n = q_c\left[\Gamma/\Lambda_n - 1\right],$$

where

$$\Gamma = 1 - \cos 2\varphi - i\sin 2\varphi\tanh[2\mu t + 2p]\ ,$$

$$\Lambda_n = 1 \pm \cos\varphi[\cos\beta]^{-1}\ \mathrm{sech}[2\mu t + 2p]\cos[2n\beta]\ ,$$

where ('+' corresponds to $\vartheta = -\beta$).

The heteroclinic orbit (3.24) represents the figure eight structure. If we denote by S the circle, we have the topological identification:

$$(\text{figure 8}) \otimes S = \bigcup_{p\in(-\infty,\infty),\ \gamma\in[0,2\pi]} Q_n(p, \gamma, a, \omega, \pm, N)\ .$$

3.3. Davey-Stewartson II (DSII) Equations

Consider the Davey-Stewartson II equations (DSII),

$$(3.25) \qquad \begin{cases} i\partial_t q = [\partial_x^2 - \partial_y^2]q + [2(|q|^2 - \omega^2) + u_y]q\,, \\[2mm] [\partial_x^2 + \partial_y^2]u = -4\partial_y|q|^2\,, \end{cases}$$

where q and u are respectively complex-valued and real-valued functions of three variables (t, x, y), and ω is a positive constant. We pose periodic boundary conditions,

$$q(t, x + L_1, y) = q(t, x, y) = q(t, x, y + L_2)\ ,$$

$$u(t, x + L_1, y) = u(t, x, y) = u(t, x, y + L_2)\ ,$$

and the even constraint,

$$q(t, -x, y) = q(t, x, y) = q(t, x, -y) \,,$$
$$u(t, -x, y) = u(t, x, y) = u(t, x, -y) \,.$$

Its Lax pair is defined as:

$$(3.26) \qquad\qquad L\psi \;=\; \lambda\psi \,,$$
$$(3.27) \qquad\qquad \partial_t \psi \;=\; A\psi \,,$$

where $\psi = (\psi_1, \psi_2)$, and

$$L = \begin{pmatrix} D^- & q \\ \bar{q} & D^+ \end{pmatrix} \,,$$

$$A = i \left[2 \begin{pmatrix} -\partial_x^2 & q\partial_x \\ \bar{q}\partial_x & \partial_x^2 \end{pmatrix} + \begin{pmatrix} r_1 & (D^+ q) \\ -(D^-\bar{q}) & r_2 \end{pmatrix} \right] \,,$$

$$(3.28) \qquad D^+ = \alpha\partial_y + \partial_x \,, \qquad D^- = \alpha\partial_y - \partial_x \,, \qquad \alpha^2 = -1 \,.$$

r_1 and r_2 have the expressions,

$$(3.29) \qquad\qquad r_1 = \frac{1}{2}[-w + iv] \,, \qquad r_2 = \frac{1}{2}[w + iv] \,,$$

where u and v are real-valued functions satisfying

$$(3.30) \qquad\qquad [\partial_x^2 + \partial_y^2]w = 2[\partial_x^2 - \partial_y^2]|q|^2 \,,$$
$$(3.31) \qquad\qquad [\partial_x^2 + \partial_y^2]v = i4\alpha\partial_x\partial_y|q|^2 \,,$$

and $w = 2(|q|^2 - \omega^2) + u_y$. Notice that DSII (3.25) is invariant under the transformation σ:

$$(3.32) \qquad\qquad \sigma \circ (q, \bar{q}, r_1, r_2; \alpha) = (q, \bar{q}, -r_2, -r_1; -\alpha) \,.$$

Applying the transformation σ (3.32) to the Lax pair (3.26, 3.27), we have a congruent Lax pair for which the compatibility condition gives the same DSII. The congruent Lax pair is given as:

$$(3.33) \qquad\qquad \hat{L}\hat{\psi} \;=\; \lambda\hat{\psi} \,,$$
$$(3.34) \qquad\qquad \partial_t \hat{\psi} \;=\; \hat{A}\hat{\psi} \,,$$

where $\hat{\psi} = (\hat{\psi}_1, \hat{\psi}_2)$, and

$$\hat{L} = \begin{pmatrix} -D^+ & q \\ \bar{q} & -D^- \end{pmatrix} \,,$$

$$\hat{A} = i \left[2 \begin{pmatrix} -\partial_x^2 & q\partial_x \\ \bar{q}\partial_x & \partial_x^2 \end{pmatrix} + \begin{pmatrix} -r_2 & -(D^-q) \\ (D^+\bar{q}) & -r_1 \end{pmatrix} \right] \,.$$

The compatibility condition of the Lax pair (3.26, 3.27),

$$\partial_t L = [A, L] \,,$$

where $[A, L] = AL - LA$, and the compatibility condition of the congruent Lax pair (3.33, 3.34),

$$\partial_t \hat{L} = [\hat{A}, \hat{L}]$$

give the same DSII (3.25). Let (q, u) be a solution to the DSII (3.25), and let λ_0 be any value of λ. Let $\psi = (\psi_1, \psi_2)$ be a solution to the Lax pair (3.26, 3.27) at $(q, \bar{q}, r_1, r_2; \lambda_0)$. Define the matrix operator:

$$\Gamma = \begin{bmatrix} \Lambda + a & b \\ c & \Lambda + d \end{bmatrix},$$

where $\Lambda = \alpha\partial_y - \lambda$, and a, b, c, d are functions defined as:

$$a = \frac{1}{\Delta}\left[\psi_2 \Lambda_2 \bar{\psi}_2 + \bar{\psi}_1 \Lambda_1 \psi_1\right],$$

$$b = \frac{1}{\Delta}\left[\bar{\psi}_2 \Lambda_1 \psi_1 - \psi_1 \Lambda_2 \bar{\psi}_2\right],$$

$$c = \frac{1}{\Delta}\left[\bar{\psi}_1 \Lambda_1 \psi_2 - \psi_2 \Lambda_2 \bar{\psi}_1\right],$$

$$d = \frac{1}{\Delta}\left[\bar{\psi}_2 \Lambda_1 \psi_2 + \psi_1 \Lambda_2 \bar{\psi}_1\right],$$

in which $\Lambda_1 = \alpha\partial_y - \lambda_0$, $\Lambda_2 = \alpha\partial_y + \bar{\lambda}_0$, and

$$\Delta = -\left[|\psi_1|^2 + |\psi_2|^2\right].$$

Define a transformation as follows:

$$\begin{cases} (q, r_1, r_2) & \to & (Q, R_1, R_2), \\ \phi & \to & \Phi; \end{cases}$$

$$Q = q - 2b,$$

(3.35) $$R_1 = r_1 + 2(D^+ a),$$

$$R_2 = r_2 - 2(D^- d),$$

$$\Phi = \Gamma\phi;$$

where ϕ is any solution to the Lax pair (3.26, 3.27) at $(q, \bar{q}, r_1, r_2; \lambda)$, D^+ and D^- are defined in (3.28), we have the following theorem [**121**].

THEOREM 3.6. *The transformation (3.35) is a Bäcklund-Darboux transformation. That is, the function Q defined through the transformation (3.35) is also a solution to the DSII (3.25). The function Φ defined through the transformation (3.35) solves the Lax pair (3.26, 3.27) at $(Q, \bar{Q}, R_1, R_2; \lambda)$.*

3.3.1. An Example . Instead of using L_1 and L_2 to describe the periods of the periodic boundary condition, one can introduce κ_1 and κ_2 as $L_1 = \frac{2\pi}{\kappa_1}$ and $L_2 = \frac{2\pi}{\kappa_2}$. Consider the spatially independent solution,

(3.36) $$q_c = \eta\exp\{-2i[\eta^2 - \omega^2]t + i\gamma\}.$$

The dispersion relation for the linearized DSII at q_c is

$$\Omega = \pm\frac{|\xi_1^2 - \xi_2^2|}{\sqrt{\xi_1^2 + \xi_2^2}}\sqrt{4\eta^2 - (\xi_1^2 + \xi_2^2)}, \quad \text{for } \delta q \sim q_c\exp\{i(\xi_1 x + \xi_2 y) + \Omega t\},$$

where $\xi_1 = k_1\kappa_1$, $\xi_2 = k_2\kappa_2$, and k_1 and k_2 are integers. We restrict κ_1 and κ_2 as follows to have only two unstable modes $(\pm\kappa_1, 0)$ and $(0, \pm\kappa_2)$,

$$\kappa_2 < \kappa_1 < 2\kappa_2 \ , \quad \kappa_1^2 < 4\eta^2 < \min\{\kappa_1^2 + \kappa_2^2, 4\kappa_2^2\} \ ,$$

or

$$\kappa_1 < \kappa_2 < 2\kappa_1 \ , \quad \kappa_2^2 < 4\eta^2 < \min\{\kappa_1^2 + \kappa_2^2, 4\kappa_1^2\} \ .$$

The Bloch eigenfunction of the Lax pair (3.26) and (3.27) is given as,

$$(3.37) \qquad \psi = c(t) \begin{bmatrix} -q_c \\ \chi \end{bmatrix} \exp\left\{ i(\xi_1 x + \xi_2 y) \right\} \ ,$$

where

$$c(t) = c_0 \exp\left\{ [2\xi_1(i\alpha\xi_2 - \lambda) + ir_2] t \right\} \ ,$$
$$r_2 - r_1 = 2(|q_c|^2 - \omega^2) \ ,$$
$$\chi = (i\alpha\xi_2 - \lambda) - i\xi_1 \ ,$$
$$(i\alpha\xi_2 - \lambda)^2 + \xi_1^2 = \eta^2 \ .$$

For the iteration of the Bäcklund-Darboux transformations, one needs two sets of eigenfunctions. First, we choose $\xi_1 = \pm\frac{1}{2}\kappa_1$, $\xi_2 = 0$, $\lambda_0 = \sqrt{\eta^2 - \frac{1}{4}\kappa_1^2}$ (for a fixed branch),

$$(3.38) \qquad \psi^\pm = c^\pm \begin{bmatrix} -q_c \\ \chi^\pm \end{bmatrix} \exp\left\{ \pm i\frac{1}{2}\kappa_1 x \right\} \ ,$$

where

$$c^\pm = c_0^\pm \exp\left\{ [\mp\kappa_1\lambda_0 + ir_2] t \right\} \ ,$$
$$\chi^\pm = -\lambda_0 \mp i\frac{1}{2}\kappa_1 = \eta e^{\mp i(\frac{\pi}{2} + \vartheta_1)} \ .$$

We apply the Bäcklund-Darboux transformations with $\psi = \psi^+ + \psi^-$, which generates the unstable foliation associated with the $(\kappa_1, 0)$ and $(-\kappa_1, 0)$ linearly unstable modes. Then, we choose $\xi_2 = \pm\frac{1}{2}\kappa_2$, $\lambda = 0$, $\xi_1^0 = \sqrt{\eta^2 - \frac{1}{4}\kappa_2^2}$ (for a fixed branch),

$$(3.39) \qquad \phi_\pm = c_\pm \begin{bmatrix} -q_c \\ \chi_\pm \end{bmatrix} \exp\left\{ i(\xi_1^0 x \pm \frac{1}{2}\kappa_2 y) \right\} \ ,$$

where

$$c_\pm = c_\pm^0 \exp\left\{ [\pm i\alpha\kappa_2\xi_1^0 + ir_2] t \right\} \ ,$$
$$\chi_\pm = \pm i\alpha\frac{1}{2}\kappa_2 - i\xi_1^0 = \pm\eta e^{\mp i\vartheta_2} \ .$$

We start from these eigenfunctions $\phi_\pm$ to generate $\Gamma\phi_\pm$ through Bäcklund-Darboux transformations, and then iterate the Bäcklund-Darboux transformations with $\Gamma\phi_+ + \Gamma\phi_-$ to generate the unstable foliation associated with all the linearly unstable modes $(\pm\kappa_1, 0)$ and $(0, \pm\kappa_2)$. It turns out that the following representations are convenient,

$$(3.40) \qquad \psi^{\pm} \;=\; \sqrt{c_0^+ c_0^-}\, e^{ir_2 t} \begin{pmatrix} v_1^{\pm} \\ v_2^{\pm} \end{pmatrix} ,$$

$$(3.41) \qquad \phi_{\pm} \;=\; \sqrt{c_+^0 c_-^0}\, e^{i\xi_1^0 x + i r_2 t} \begin{pmatrix} w_1^{\pm} \\ w_2^{\pm} \end{pmatrix} ,$$

where

$$v_1^{\pm} = -q_c e^{\mp \frac{\tau}{2} \pm i\tilde{x}} , \quad v_2^{\pm} = \eta e^{\mp \frac{\tau}{2} \pm i\tilde{z}} ,$$

$$w_1^{\pm} = -q_c e^{\pm \frac{\hat{\tau}}{2} \pm i\hat{y}} , \quad w_2^{\pm} = \pm\eta e^{\pm \frac{\hat{\tau}}{2} \pm i\hat{z}} ,$$

and

$$c_0^+/c_0^- = e^{\rho + i\vartheta} , \quad \tau = 2\kappa_1 \lambda_0 t - \rho , \quad \tilde{x} = \frac{1}{2}\kappa_1 x + \frac{\vartheta}{2} , \tilde{z} = \tilde{x} - \frac{\pi}{2} - \vartheta_1 ,$$

$$c_+^0/c_-^0 = e^{\hat{\rho} + i\hat{\vartheta}} , \quad \hat{\tau} = 2i\alpha\kappa_2 \xi_1^0 t + \hat{\rho} , \quad \hat{y} = \frac{1}{2}\kappa_2 y + \frac{\hat{\vartheta}}{2} , \hat{z} = \hat{y} - \vartheta_2 .$$

The following representations are also very useful,

$$(3.42) \qquad \psi \;=\; \psi^+ + \psi^- = 2\sqrt{c_0^+ c_0^-}\, e^{ir_2 t} \begin{pmatrix} v_1 \\ v_2 \end{pmatrix} ,$$

$$(3.43) \qquad \phi \;=\; \phi^+ + \phi^- = 2\sqrt{c_+^0 c_-^0}\, e^{i\xi_1^0 x + i r_2 t} \begin{pmatrix} w_1 \\ w_2 \end{pmatrix} ,$$

where

$$v_1 = -q_c[\cosh \frac{\tau}{2} \cos \tilde{x} - i \sinh \frac{\tau}{2} \sin \tilde{x}] , \quad v_2 = \eta[\cosh \frac{\tau}{2} \cos \tilde{z} - i \sinh \frac{\tau}{2} \sin \tilde{z}] ,$$

$$w_1 = -q_c[\cosh \frac{\hat{\tau}}{2} \cos \hat{y} + i \sinh \frac{\hat{\tau}}{2} \sin \hat{y}] , \quad w_2 = \eta[\sinh \frac{\hat{\tau}}{2} \cos \hat{z} + i \cosh \frac{\hat{\tau}}{2} \sin \hat{z}] .$$

Applying the Bäcklund-Darboux transformations (3.35) with ψ given in (3.42), we have the representations,

$$(3.44) \quad a \;=\; -\lambda_0 \ \mathrm{sech}\ \tau \sin(\tilde{x} + \tilde{z}) \sin(\tilde{x} - \tilde{z})$$
$$\times \left[1 + \ \mathrm{sech}\ \tau \cos(\tilde{x} + \tilde{z}) \cos(\tilde{x} - \tilde{z}) \right]^{-1} ,$$

$$(3.45) \quad b \;=\; -q_c\tilde{b} = -\frac{\lambda_0 q_c}{\eta} \left[\cos(\tilde{x} - \tilde{z}) - i \tanh \tau \sin(\tilde{x} - \tilde{z}) \right.$$
$$\left. + \ \mathrm{sech}\ \tau \cos(\tilde{x} + \tilde{z}) \right] \left[1 + \ \mathrm{sech}\ \tau \cos(\tilde{x} + \tilde{z}) \cos(\tilde{x} - \tilde{z}) \right]^{-1} ,$$

$$(3.46) \qquad c = \bar{b} , \quad d = -\bar{a} = -a .$$

The evenness of b in x is enforced by the requirement that $\vartheta - \vartheta_1 = \pm\frac{\pi}{2}$, and

$$(3.47) \qquad a^{\pm} = \mp\lambda_0 \,\mathrm{sech}\, \tau \cos\vartheta_1 \sin(\kappa_1 x)$$
$$\times \left[1 \mp \,\mathrm{sech}\, \tau \sin\vartheta_1 \cos(\kappa_1 x) \right]^{-1} ,$$

$$(3.48) \qquad b^{\pm} = -q_c \tilde{b}^{\pm} = -\frac{\lambda_0 q_c}{\eta}\left[-\sin\vartheta_1 - i \tanh\tau \cos\vartheta_1 \right.$$
$$\left. \pm \,\mathrm{sech}\, \tau \cos(\kappa_1 x) \right]\left[1 \mp \,\mathrm{sech}\, \tau \sin\vartheta_1 \cos(\kappa_1 x) \right]^{-1} ,$$

$$(3.49) \qquad c = \bar{b} , \quad d = -\bar{a} = -a .$$

Notice also that $a^{\pm}$ is an odd function in x. Under the above Bäcklund-Darboux transformations, the eigenfunctions $\phi_{\pm}$ (3.39) are transformed into

$$(3.50) \qquad\qquad \varphi^{\pm} = \Gamma\phi_{\pm} ,$$

where

$$\Gamma = \begin{bmatrix} \Lambda + a & b \\ \bar{b} & \Lambda - a \end{bmatrix} ,$$

and $\Lambda = \alpha\partial_y - \lambda$ with λ evaluated at 0. Let $\varphi = \varphi^+ + \varphi^-$ (the arbitrary constants $c_{\pm}^0$ are already included in $\varphi^{\pm}$), φ has the representation,

$$(3.51) \qquad\qquad \varphi = 2\sqrt{c_+^0 c_-^0}\, e^{i\xi_1^0 x + i r_2 t} \begin{bmatrix} -q_c W_1 \\ \eta W_2 \end{bmatrix} ,$$

where

$$W_1 = (\alpha\partial_y w_1) + a w_1 + \eta\tilde{b} w_2 , \quad W_2 = (\alpha\partial_y w_2) - a w_2 + \eta\bar{\tilde{b}} w_1 .$$

We generate the coefficients in the Bäcklund-Darboux transformations (3.35) with φ (the iteration of the Bäcklund-Darboux transformations),

$$(3.52) \qquad a^{(I)} = -\left[W_2(\alpha\partial_y \overline{W_2}) + \overline{W_1}(\alpha\partial_y W_1) \right]\left[|W_1|^2 + |W_2|^2 \right]^{-1} ,$$

$$(3.53) \qquad b^{(I)} = \frac{q_c}{\eta}\left[\overline{W_2}(\alpha\partial_y W_1) - W_1(\alpha\partial_y \overline{W_2}) \right]\left[|W_1|^2 + |W_2|^2 \right]^{-1} ,$$

$$(3.54) \qquad\qquad c^{(I)} = \overline{b^{(I)}} , \quad d^{(I)} = -\overline{a^{(I)}} ,$$

where

$$W_2(\alpha\partial_y\overline{W_2}) + \overline{W_1}(\alpha\partial_y W_1)$$

$$= \frac{1}{2}\alpha\kappa_2\left\{\cosh\hat{\tau}\left[-\alpha\kappa_2 a + ia\eta(\tilde{b} + \bar{\tilde{b}})\cos\vartheta_2\right]\right.$$

$$+ \left[\frac{1}{4}\kappa_2^2 - a^2 - \eta^2|\tilde{b}|^2\right]\cos(\hat{y} + \hat{z})\sin\vartheta_2 + \left.\sinh\hat{\tau}\left[a\eta(\tilde{b} - \bar{\tilde{b}})\sin\vartheta_2\right]\right\},$$

$$|W_1|^2 + |W_2|^2$$

$$= \cosh\hat{\tau}\left[a^2 + \frac{1}{4}\kappa_2^2 + \eta^2|\tilde{b}|^2 + ia\kappa_2\eta\frac{1}{2}(\tilde{b} + \bar{\tilde{b}})\cos\vartheta_2\right]$$

$$+ \left[\frac{1}{4}\kappa_2^2 - a^2 - \eta^2|\tilde{b}|^2\right]\sin(\hat{y} + \hat{z})\sin\vartheta_2 + \sinh\hat{\tau}\left[\alpha\kappa_2\eta\frac{1}{2}(\tilde{b} - \bar{\tilde{b}})\sin\vartheta_2\right],$$

$$\overline{W_2}(\alpha\partial_y W_1) - W_1(\alpha\partial_y\overline{W_2})$$

$$= \frac{1}{2}\alpha\kappa_2\left\{\cosh\hat{\tau}\left[-\alpha\kappa_2\eta\tilde{b} + i(-a^2 + \frac{1}{4}\kappa_2^2 + \eta^2\tilde{b}^2)\cos\vartheta_2\right]\right.$$

$$+ \left.\sinh\hat{\tau}(a^2 - \frac{1}{4}\kappa_2^2 + \eta^2\tilde{b}^2)\sin\vartheta_2\right\}.$$

The new solution to the focusing Davey-Stewartson II equation (3.25) is given by

$$(3.55) \qquad\qquad Q = q_c - 2b - 2b^{(I)}.$$

The evenness of $b^{(I)}$ in y is enforced by the requirement that $\hat{\vartheta} - \vartheta_2 = \pm\frac{\pi}{2}$. In fact, we have

LEMMA 3.7. *Under the requirements that $\vartheta - \vartheta_1 = \pm\frac{\pi}{2}$, and $\hat{\vartheta} - \vartheta_2 = \pm\frac{\pi}{2}$,*

$$(3.56) \qquad b(-x) = b(x), \quad b^{(I)}(-x, y) = b^{(I)}(x, y) = b^{(I)}(x, -y),$$

and $Q = q_c - 2b - 2b^{(I)}$ is even in both x and y.

Proof: It is a direct verification by noticing that under the requirements, a is an odd function in x. Q.E.D.

The asymptotic behavior of Q can be computed directly. In fact, we have the asymptotic phase shift lemma.

LEMMA 3.8 (Asymptotic Phase Shift Lemma). *For $\lambda_0 > 0$, $\xi_1^0 > 0$, and $i\alpha = 1$, as $t \to \pm\infty$,*

$$(3.57) \qquad\qquad Q = q_c - 2b - 2b^{(I)} \to q_c e^{i\pi} e^{\mp i2(\vartheta_1 - \vartheta_2)}.$$

In comparison, the asymptotic phase shift of the first application of the Bäcklund-Darboux transformations is given by

$$q_c - 2b \to q_c e^{\mp i2\vartheta_1}.$$

3.4. Other Soliton Equations

In general, one can classify soliton equations into two categories. Category I consists of those equations possessing instabilities, under periodic boundary condition. In their phase space, figure-eight structures (i.e. separatrices) exist. Category

II consists of those equations possessing no instability, under periodic boundary condition. In their phase space, no figure-eight structure (i.e. separatrix) exists. Typical Category I soliton equations are, for example, focusing nonlinear Schrödinger equation, sine-Gordon equation [127], modified KdV equation. Typical Category II soliton equations are, for example, KdV equation, defocusing nonlinear Schrödinger equation, sinh-Gordon equation, Toda lattice. In principle, figure-eight structures for Category I soliton equations can be constructed through Bäcklund-Darboux transformations, as illustrated in previous sections. It should be remarked that Bäcklund-Darboux transformations still exist for Category II soliton equations, but do not produce any figure-eight structure. A good reference on Bäcklund-Darboux transformations is [152].

CHAPTER 4

Melnikov Vectors

4.1. 1D Cubic Nonlinear Schrödinger Equation

We select the NLS (3.1) as our first example to show how to establish Melnikov vectors. We continue from Section 3.1.

DEFINITION 4.1. Define the sequence of functionals F_j as follows,

$$(4.1) \qquad F_j(\vec{q}) = \Delta(\lambda_j^c(\vec{q}), \vec{q}),$$

where λ_j^c's are the critical points, $\vec{q} = (q, -\bar{q})$.

We have the lemma [137]:

LEMMA 4.2. *If $\lambda_j^c(\vec{q})$ is a simple critical point of $\Delta(\lambda)$ [i.e., $\Delta''(\lambda_j^c) \neq 0$], F_j is analytic in a neighborhood of $\vec{q}$, with first derivative given by*

$$(4.2) \qquad \frac{\delta F_j}{\delta q} = \frac{\delta \Delta}{\delta q}\bigg|_{\lambda=\lambda_j^c} + \frac{\partial \Delta}{\partial \lambda}\bigg|_{\lambda=\lambda_j^c} \frac{\delta \lambda_j^c}{\delta q} = \frac{\delta \Delta}{\delta q}\bigg|_{\lambda=\lambda_j^c} ,$$

where

$$(4.3) \qquad \frac{\delta}{\delta \vec{q}} \Delta(\lambda; \vec{q}) = i \frac{\sqrt{\Delta^2 - 4}}{W[\psi^+, \psi^-]} \left[\begin{array}{c} \psi_2^+(x; \lambda)\psi_2^-(x; \lambda) \\[2mm] \psi_1^+(x; \lambda)\psi_1^-(x; \lambda) \end{array} \right] ,$$

and the Bloch eigenfunctions $\psi^{\pm}$ have the property that

$$(4.4) \qquad \psi^{\pm}(x + 2\pi; \lambda) = \rho^{\pm 1}\psi^{\pm}(x; \lambda) ,$$

for some ρ, also the Wronskian is given by

$$W[\psi^+, \psi^-] = \psi_1^+\psi_2^- - \psi_2^+\psi_1^- .$$

In addition, Δ' is given by

$$(4.5) \qquad \frac{d\Delta}{d\lambda} = -i \frac{\sqrt{\Delta^2 - 4}}{W[\psi^+, \psi^-]} \int_0^{2\pi} [\psi_1^+\psi_2^- + \psi_2^+\psi_1^-] \, dx .$$

Proof: To prove this lemma, one calculates using variation of parameters:

$$\delta M(x; \lambda) = M(x) \int_0^x M^{-1}(x')\delta\hat{Q}(x')M(x')dx',$$

where

$$\delta\hat{Q} \equiv i \begin{pmatrix} 0 & \delta q \\ \delta\bar{q} & 0 \end{pmatrix} .$$

27

Thus; one obtains the formula

$$\delta\Delta(\lambda;\vec{q}) = \text{trace}\left[M(2\pi)\int_0^{2\pi} M^{-1}(x')\,\delta\hat{Q}(x')\,M(x')dx'\right],$$

which gives

$$\frac{\delta\Delta(\lambda)}{\delta q(x)} = i\,\text{trace}\left[M^{-1}(x)\begin{pmatrix} 0 & 1 \\ 0 & 0 \end{pmatrix}M(x)M(2\pi)\right],$$

(4.6)

$$\frac{\delta\Delta(\lambda)}{\delta\bar{q}(x)} = i\,\text{trace}\left[M^{-1}(x)\begin{pmatrix} 0 & 0 \\ 1 & 0 \end{pmatrix}M(x)M(2\pi)\right].$$

Next, we use the Bloch eigenfunctions $\{\psi^{\pm}\}$ to form the matrix

$$N(x;\lambda) = \begin{pmatrix} \psi_1^+ & \psi_1^- \\ \psi_2^+ & \psi_2^- \end{pmatrix}.$$

Clearly,

$$N(x;\lambda) = M(x;\lambda)N(0;\lambda)\;;$$

or equivalently,

(4.7)
$$M(x;\lambda) = N(x;\lambda)[N(0;\lambda)]^{-1}\;.$$

Since $\psi^{\pm}$ are Bloch eigenfunctions, one also has

$$N(x+2\pi;\lambda) = N(x;\lambda)\begin{pmatrix} \rho & 0 \\ 0 & \rho^{-1} \end{pmatrix},$$

which implies

$$N(2\pi;\lambda) = M(2\pi;\lambda)N(0;\lambda) = N(0;\lambda)\begin{pmatrix} \rho & 0 \\ 0 & \rho^{-1} \end{pmatrix},$$

that is,

(4.8)
$$M(2\pi;\lambda) = N(0;\lambda)\begin{pmatrix} \rho & 0 \\ 0 & \rho^{-1} \end{pmatrix}[N(0;\lambda)]^{-1}\;.$$

For any 2×2 matrix σ, equations (4.7) and (4.8) imply

$$\text{trace}\,\{[M(x)]^{-1}\,\sigma\,M(x)\,M(2\pi)\} = \text{trace}\,\{[N(x)]^{-1}\,\sigma\,N(x)\,diag(\rho,\rho^{-1})\},$$

which, through an explicit evaluation of (4.6), proves formula (4.3). Formula (4.5) is established similarly. These formulas, together with the fact that $\lambda^c(\vec{q})$ is differentiable because it is a simple zero of Δ', provide the representation of $\frac{\delta F_j}{\delta\bar{q}}$. Q.E.D.

REMARK 4.3. Formula (4.2) for the $\frac{\delta F_j}{\delta\bar{q}}$ is actually valid even if $\Delta''(\lambda_j^c(\vec{q});\vec{q}) = 0$. Consider a function $\vec{q}_*$ at which

$$\Delta''(\lambda_j^c(\vec{q}_*);\vec{q}_*) = 0,$$

and thus at which $\lambda_j^c(\vec{q})$ fails to be analytic. For $\vec{q}$ near $\vec{q}_*$, one has

$$\Delta'(\lambda_j^c(\vec{q});\vec{q}) = 0;$$

$$\frac{\delta}{\delta\vec{q}}\Delta'(\lambda_j^c(\vec{q});\vec{q}) = \Delta''\frac{\delta}{\delta\vec{q}}\lambda_j^c + \frac{\delta}{\delta\vec{q}}\Delta' = 0;$$

that is,

$$\frac{\delta}{\delta\vec{q}}\lambda_j^c = -\frac{1}{\Delta''}\frac{\delta}{\delta\vec{q}}\Delta' \ .$$

Thus,

$$\frac{\delta}{\delta\vec{q}}F_j = \frac{\delta}{\delta\vec{q}}\Delta + \Delta'\frac{\delta}{\delta\vec{q}}\lambda_j^c = \frac{\delta}{\delta\vec{q}}\Delta - \frac{\Delta'}{\Delta''}\frac{\delta}{\delta\vec{q}}\Delta' \ |_{\lambda=\lambda_j^c(\vec{q})} \ .$$

Since $\frac{\Delta'}{\Delta''} \to 0$, as $\vec{q} \to \vec{q}_*$, one still has formula (4.2) at $\vec{q} = \vec{q}_*$:

$$\frac{\delta}{\delta\vec{q}}F_j = \frac{\delta}{\delta\vec{q}}\Delta \ |_{\lambda=\lambda_j^c(\vec{q})} \ .$$

The NLS (3.1) is a Hamiltonian system:

$$(4.9) \qquad\qquad iq_t = \frac{\delta H}{\delta\bar{q}} \ ,$$

where

$$H = \int_0^{2\pi} \{-|q_x|^2 + |q|^4\} \ dx \ .$$

COROLLARY 4.4. *For any fixed* $\lambda \in \mathbb{C}$, $\Delta(\lambda,\vec{q})$ *is a constant of motion of the NLS (3.1). In fact,*

$$\{\Delta(\lambda,\vec{q}), H(\vec{q})\} = 0 \ , \quad \{\Delta(\lambda,\vec{q}), \Delta(\lambda',\vec{q})\} = 0 \ , \quad \forall \lambda, \lambda' \in \mathbb{C} \ ,$$

where for any two functionals E *and* F, *their Poisson bracket is defined as*

$$\{E, F\} = \int_0^{2\pi} \left[\frac{\delta E}{\delta q}\frac{\delta F}{\delta\bar{q}} - \frac{\delta E}{\delta\bar{q}}\frac{\delta F}{\delta q} \right] dx \ .$$

Proof: The corollary follows from a direction calculation from the spatial part (3.2) of the Lax pair and the representation (4.3). Q.E.D.

For each fixed $\vec{q}$, Δ is an entire function of λ; therefore, can be determined by its values at a countable number of values of λ. The invariance of Δ characterizes the isospectral nature of the NLS equation.

COROLLARY 4.5. *The functionals* F_j *are constants of motion of the NLS (3.1). Their gradients provide Melnikov vectors:*

$$(4.10) \qquad grad \ F_j(\vec{q}) = i\frac{\sqrt{\Delta^2 - 4}}{W[\psi^+, \psi^-]} \left[\begin{array}{c} \psi_2^+(x; \lambda_j^c)\psi_2^-(x; \lambda_j^c) \\ \psi_1^+(x; \lambda_j^c)\psi_1^-(x; \lambda_j^c) \end{array} \right] \ .$$

The distribution of the critical points λ_j^c are described by the following counting lemma [137],

LEMMA 4.6 (Counting Lemma for Critical Points). *For* $q \in H^1$, *set* $N = N(\|q\|_1) \in \mathbb{Z}^+$ *by*

$$N(\|q\|_1) = 2\left[\|q\|_0^2 \cosh\left(\|q\|_0\right) + 3\|q\|_1 \sinh\left(\|q\|_0\right)\right],$$

where $[x] = $ *first integer greater than* x. *Consider*

$$\Delta'(\lambda; \vec{q}) = \frac{d}{d\lambda} \Delta(\lambda; \vec{q}).$$

Then

(1) $\Delta'(\lambda; \vec{q})$ *has exactly* $2N + 1$ *zeros (counted according to multiplicity) in the interior of the disc* $D = \{\lambda \in \mathbb{C} : |\lambda| < (2N + 1)\frac{\pi}{2}\}$;

(2) $\forall k \in \mathbb{Z}, |k| > N, \Delta'(\lambda, \vec{q})$ *has exactly one zero in each disc*
$\{\lambda \in \mathbb{C}: \ |\lambda - k\pi| < \frac{\pi}{4}\}$.

(3) $\Delta'(\lambda; \vec{q})$ *has no other zeros.*

(4) *For* $|\lambda| > (2N + 1)\frac{\pi}{2}$, *the zeros of* Δ', $\{\lambda_j^c, |j| > N\}$, *are all real, simple, and satisfy the asymptotics*

$$\lambda_j^c = j\pi + o(1) \quad \text{as } |j| \to \infty.$$

4.1.1. Melnikov Integrals. When studying perturbed integrable systems, the figure-eight structures often lead to chaotic dynamics through homoclinic bifurcations. An extremely powerful tool for detecting homoclinic orbits is the so-called Melnikov integral method [158], which uses "Melnikov integrals" to provide estimates of the distance between the center-unstable manifold and the center-stable manifold of a normally hyperbolic invariant manifold. The Melnikov integrals are often integrals in time of the inner products of certain Melnikov vectors with the perturbations in the perturbed integrable systems. This implies that the Melnikov vectors play a key role in the Melnikov integral method. First, we consider the case of one unstable mode associated with a complex double point ν, for which the homoclinic orbit is given by Bäcklund-Darboux formula (3.7),

$$Q(x,t) \equiv q(x,t) \ + \ 2(\nu - \bar{\nu}) \ \frac{\phi_1 \bar{\phi}_2}{\phi_1 \bar{\phi}_1 + \phi_2 \bar{\phi}_2} \ ,$$

where q lies in a normally hyperbolic invariant manifold and ϕ denotes a general solution to the Lax pair (3.2, 3.3) at (q, ν), $\phi = c_+\phi^+ + c_-\phi^-$, and $\phi^\pm$ are Bloch eigenfunctions. Next, we consider the perturbed NLS,

$$iq_t = q_{xx} + 2|q|^2 q + i\epsilon f(q, \bar{q}) \ ,$$

where ϵ is the perturbation parameter. The *Melnikov integral* can be defined using the constant of motion F_j, where $\lambda_j^c = \nu$ [137]:

$$(4.11) \qquad M_j \equiv \int_{-\infty}^{+\infty} \int_0^{2\pi} \{\frac{\delta F_j}{\delta q} f + \frac{\delta F_j}{\delta \bar{q}} \bar{f}\}_{q=Q} \ dxdt \ ,$$

where the integrand is evaluated along the unperturbed homoclinic orbit $q = Q$, and the Melnikov vector $\frac{\delta F_j}{\delta \bar{q}}$ has been given in the last section, which can be expressed rather explicitly using the Bäcklund-Darboux transformation [137] . We begin with the expression (4.10),

$$(4.12) \qquad \frac{\delta F_j}{\delta \vec{q}} = i \frac{\sqrt{\Delta^2 - 4}}{W[\Phi^+, \Phi^-]} \left(\begin{array}{c} \Phi_2^+ \Phi_2^- \\ \Phi_1^+ \Phi_1^- \end{array} \right)$$

where $\Phi^\pm$ are Bloch eigenfunctions at (Q, ν), which can be obtained from Bäcklund-Darboux formula (3.7):

$$\Phi^\pm(x, t; \nu) \ \equiv \ G(\nu; \nu; \phi) \ \phi^\pm(x, t; \nu) \ ,$$

with the transformation matrix G given by

$$G = G(\lambda; \nu; \phi) = N \left(\begin{array}{cc} \lambda - \nu & 0 \\ 0 & \lambda - \bar{\nu} \end{array} \right) N^{-1},$$

$$N \equiv \left[\begin{array}{cc} \phi_1 & -\bar{\phi}_2 \\ \phi_2 & \bar{\phi}_1 \end{array} \right] .$$

These Bäcklund-Darboux formulas are rather easy to manipulate to obtain explicit information. For example, the transformation matrix $G(\lambda, \nu)$ has a simple limit as $\lambda \to \nu$:

$$(4.13) \qquad \lim_{\lambda \to \nu} G(\lambda, \nu) = \frac{\nu - \bar{\nu}}{|\phi|^2} \begin{pmatrix} \phi_2 \bar{\phi}_2 & -\phi_1 \bar{\phi}_2 \\ -\phi_2 \bar{\phi}_1 & \phi_1 \bar{\phi}_1 \end{pmatrix}$$

where $|\phi|^2$ is defined by

$$|\vec{\phi}|^2 \equiv \phi_1 \bar{\phi}_1 + \phi_2 \bar{\phi}_2.$$

With formula (4.13) one quickly calculates

$$\Phi^\pm = \pm c_\mp \; W[\phi^+, \phi^-] \; \frac{\nu - \bar{\nu}}{|\phi|^2} \begin{pmatrix} \bar{\phi}_2 \\ -\bar{\phi}_1 \end{pmatrix}$$

from which one sees that Φ^+ and Φ^- are linearly dependent at (Q, ν),

$$\Phi^+ = -\frac{c_-}{c_+} \; \Phi^-.$$

REMARK 4.7. For Q on the figure-eight, the two Bloch eigenfunctions $\Phi^\pm$ are linearly dependent. Thus, the geometric multiplicity of ν is only one, even though its algebraic multiplicity is two or higher.

Using L'Hospital's rule, one gets

$$(4.14) \qquad \frac{\sqrt{\Delta^2 - 4}}{W[\Phi^+, \Phi^-]} = \frac{\sqrt{\Delta(\nu)\Delta''(\nu)}}{(\nu - \bar{\nu}) \; W[\phi^+, \phi^-]} \;.$$

With formulas (4.12, 4.13, 4.14), one obtains the explicit representation of the grad F_j [137]:

$$(4.15) \qquad \frac{\delta F_j}{\delta \vec{q}} = C_\nu \frac{c_+ c_- W[\psi^{(+)}, \psi^{(-)}]}{|\phi|^4} \begin{pmatrix} \bar{\phi}_2^2 \\ -\bar{\phi}_1^2 \end{pmatrix},$$

where the constant C_ν is given by

$$C_\nu \equiv i(\nu - \bar{\nu}) \; \sqrt{\Delta(\nu)\Delta''(\nu)} \;.$$

With these ingredients, one obtains the following beautiful representation of the Melnikov function associated to the general complex double point ν [137]:

$$(4.16) \quad M_j = C_\nu \, c_+ c_- \int_{-\infty}^{+\infty} \int_0^{2\pi} W[\phi^+, \phi^-] \left[\frac{(\bar{\phi}_2^2)f(Q, \bar{Q}) + (\bar{\phi}_1^2)\overline{f(Q, \bar{Q})}}{|\phi|^4} \right] dx dt.$$

In the case of several complex double points, each associated with an instability, one can iterate the Bäcklund-Darboux tranformations and use those functionals F_j which are associated with each complex double point to obtain representations *Melnikov Vectors*. In general, the relation between $\frac{\delta F_j}{\delta \vec{q}}$ and double points can be summarized in the following lemma [137],

LEMMA 4.8. *Except for the trivial case* $q = 0$,

$$(a). \qquad \frac{\delta F_j}{\delta q} = 0 \Leftrightarrow \frac{\delta F_j}{\delta \bar{q}} = 0 \Leftrightarrow M(2\pi, \lambda_j^c; \; \vec{q}_*) = \pm I.$$

$$(b). \qquad \frac{\delta F_j}{\delta \vec{q}} \Big|_{\vec{q}_*} = 0 \Rightarrow \Delta'(\lambda_j^c(\vec{q}_*); \; \vec{q}_*) = 0, \Rightarrow |F_j(\vec{q}_*)| = 2,$$

$$\Rightarrow \lambda_j^c(\vec{q}_*) \; \textit{is a multiple point,}$$

where I is the 2×2 identity matrix.

The Bäcklund-Darboux transformation theorem indicates that the figure-eight structure is attached to a complex double point. The above lemma shows that at the origin of the figure-eight, the gradient of F_j vanishes. Together they indicate that the the gradient of F_j along the figure-eight is a perfect Melnikov vector.

Example: When $\frac{1}{2} < c < 1$ in (3.9), and choosing $\vartheta = \pi$ in (3.14), one can get the Melnikov vector field along the homoclinic orbit (3.14),

$$(4.17) \qquad \frac{\delta F_1}{\delta q} = 2\pi \, \sin^2 p \, \operatorname{sech}\tau \, \frac{[(-\sin p + i \cos p \tanh \tau) \cos x + \operatorname{sech}\tau]}{[1 - \sin p \, \operatorname{sech} \tau \cos x]^2} \, c \, e^{i\theta},$$

$$\frac{\delta F_1}{\delta \bar{q}} = \overline{\frac{\delta F_1}{\delta q}} \, .$$

4.2. Discrete Cubic Nonlinear Schrödinger Equation

The discrete cubic nonlinear Schrödinger equation (3.16) can be written in the Hamiltonian form [2] [118]:

$$(4.18) \qquad\qquad\qquad i\dot{q}_n = \rho_n \partial H / \partial \bar{q}_n \, ,$$

where $\rho_n = 1 + h^2 |q_n|^2$, and

$$H = \frac{1}{h^2} \sum_{n=0}^{N-1} \left\{ \bar{q}_n (q_{n+1} + q_{n-1}) - \frac{2}{h^2}(1 + \omega^2 h^2) \ln \rho_n \right\} \, .$$

$\sum_{n=0}^{N-1} \left\{ \bar{q}_n (q_{n+1} + q_{n-1}) \right\}$ itself is also a constant of motion. This invariant, together with H, implies that $\sum_{n=0}^{N-1} \ln \rho_n$ is a constant of motion too. Therefore,

$$(4.19) \qquad\qquad\qquad D^2 \equiv \prod_{n=0}^{N-1} \rho_n$$

is a constant of motion. We continue from Section 3.2. Using D, one can define a normalized Floquet discriminant $\tilde{\Delta}$ as

$$\tilde{\Delta} = \Delta / D \, .$$

DEFINITION 4.9. The sequence of invariants $\tilde{F}_j$ is defined as:

$$(4.20) \qquad\qquad\qquad \tilde{F}_j(\vec{q}) = \tilde{\Delta}(z_j^{(c)}(\vec{q}); \vec{q}) \, ,$$

where $\vec{q} = (q, -\bar{q})$, $q = (q_0, q_1, \cdots, q_{N-1})$.

These invariants $\tilde{F}_j$'s are perfect candidate for building Melnikov functions. The Melnikov vectors are given by the gradients of these invariants.

LEMMA 4.10. *Let $z_j^{(c)}(\vec{q})$ be a simple critical point; then*

$$(4.21) \qquad\qquad\qquad \frac{\delta \tilde{F}_j}{\delta \vec{q}_n}(\vec{q}) = \frac{\delta \tilde{\Delta}}{\delta \vec{q}_n}(z_j^{(c)}(\vec{q}); \vec{q}) \, .$$

$$(4.22) \qquad \frac{\delta \tilde{\Delta}}{\delta \vec{q}_n}(z; \vec{q}) = \frac{ih(\zeta - \zeta^{-1})}{2W_{n+1}} \begin{pmatrix} \psi_{n+1}^{(+,2)} \psi_n^{(-,2)} + \psi_n^{(+,2)} \psi_{n+1}^{(-,2)} \\[6pt] \psi_{n+1}^{(+,1)} \psi_n^{(-,1)} + \psi_n^{(+,1)} \psi_{n+1}^{(-,1)} \end{pmatrix} \, ,$$

where $\psi_n^\pm = (\psi_n^{(\pm,1)}, \psi_n^{(\pm,2)})^T$ are two Bloch functions of the Lax pair (3.2,3.3), such that

$$\psi_n^\pm = D^{n/N} \zeta^{\pm n/N} \tilde\psi_n^\pm \ ,$$

where $\tilde\psi_n^\pm$ are periodic in n with period N, $W_n = \ \det(\psi_n^+, \psi_n^-)$.

For $z_j^{(c)} = z^{(d)}$, the Melnikov vector field located on the heteroclinic orbit (3.19) is given by

$$(4.23) \qquad \frac{\delta \tilde F_j}{\delta \vec Q_n} = K \frac{W_n}{E_n A_{n+1}} \left(\begin{array}{c} [z^{(d)}]^{-2} \, \overline{\phi_n^{(1)}} \, \overline{\phi_{n+1}^{(1)}} \\[2mm] [\overline{z^{(d)}}]^{-2} \, \overline{\phi_n^{(2)}} \, \overline{\phi_{n+1}^{(2)}} \end{array} \right) \ ,$$

where $\vec Q_n = (Q_n, -\bar Q_n)$,

$$\phi_n = (\phi_n^{(1)}, \phi_n^{(2)})^T = c_+ \psi_n^+ + c_- \psi_n^- \ ,$$

$$W_n = \left| \ \psi_n^+ \ \ \psi_n^- \ \right| \ ,$$

$$E_n = |\phi_n^{(1)}|^2 + |z^{(d)}|^2 |\phi_n^{(2)}|^2 \ ,$$

$$A_n = |\phi_n^{(2)}|^2 + |z^{(d)}|^2 |\phi_n^{(1)}|^2 \ ,$$

$$K = -\frac{ih c_+ c_-}{2} |z^{(d)}|^4 (|z^{(d)}|^4 - 1) [\overline{z^{(d)}}]^{-1} \sqrt{\tilde\Delta(z^{(d)}; \vec q) \tilde\Delta''(z^{(d)}; \vec q)} \ ,$$

where

$$\tilde\Delta''(z^{(d)}; \vec q) = \frac{\partial^2 \tilde\Delta(z^{(d)}; \vec q)}{\partial z^2} \ .$$

Example: The Melnikov vector evaluated on the heteroclinic orbit (3.23) is given by

$$(4.24) \qquad \frac{\delta \tilde F_1}{\delta \vec Q_n} = \hat K \left[\hat E_n \hat A_{n+1} \right]^{-1} \operatorname{sech}[2\mu t + 2p] \left(\begin{array}{c} \hat X_n^{(1)} \\[1mm] -\hat X_n^{(2)} \end{array} \right) \ ,$$

where

$$\hat X_n^{(1)} = \Big[\cos\beta \ \operatorname{sech}[2\mu t + 2p] + \cos[(2n+1)\beta + \vartheta + \varphi]$$

$$-i \tanh[2\mu t + 2p] \sin[(2n+1)\beta + \vartheta + \varphi] \Big] e^{i2\theta(t)} \ ,$$

$$\hat X_n^{(2)} = \Big[\cos\beta \ \operatorname{sech}[2\mu t + 2p] + \cos[(2n+1)\beta + \vartheta - \varphi]$$

$$-i \tanh[2\mu t + 2p] \sin[(2n+1)\beta + \vartheta - \varphi] \Big] e^{-i2\theta(t)} \ ,$$

$$\hat K = -2N h^2 a (1 - z^4)[8\rho^{3/2} z^2]^{-1} \sqrt{\rho \cos^2\beta - 1} \ .$$

Under the even constraint, the Melnikov vector evaluated on the heteroclinic orbit (3.24) is given by

$$(4.25) \qquad \left. \frac{\delta \tilde F_1}{\delta \vec Q_n} \right|_{\text{even}} = \hat K^{(e)} \operatorname{sech}[2\mu t + 2p][\Pi_n]^{-1} \left(\begin{array}{c} \hat X_n^{(1,e)} \\[1mm] -\hat X_n^{(2,e)} \end{array} \right) \ ,$$

where

$$\hat K^{(e)} = -2N(1 - z^4)[8 a \rho^{3/2} z^2]^{-1} \sqrt{\rho \cos^2\beta - 1} \ ,$$

$$\Pi_n = \left[\cos\beta \pm \cos\varphi\,\mathrm{sech}[2\mu t + 2p]\cos[2(n-1)\beta]\right] \times$$

$$\left[\cos\beta \pm \cos\varphi\,\mathrm{sech}[2\mu t + 2p]\cos[2(n+1)\beta]\right],$$

$$\hat{X}_n^{(1,e)} = \left[\cos\beta\,\mathrm{sech}\,[2\mu t + 2p] \pm (\cos\varphi\right.$$

$$\left. -i\sin\varphi\tanh[2\mu t + 2p])\cos[2n\beta]\right]e^{i2\theta(t)},$$

$$\hat{X}_n^{(2,e)} = \left[\cos\beta\,\mathrm{sech}\,[2\mu t + 2p] \pm (\cos\varphi\right.$$

$$\left. +i\sin\varphi\tanh[2\mu t + 2p])\cos[2n\beta]\right]e^{-i2\theta(t)}.$$

4.3. Davey-Stewartson II Equations

The DSII (3.25) can be written in the Hamiltonian form,

$$(4.26) \qquad \begin{cases} iq_t &= \delta H/\delta\bar{q}\,, \\ i\bar{q}_t &= -\delta H/\delta q\,, \end{cases}$$

where

$$H = \int_0^{2\pi}\int_0^{2\pi} [|q_y|^2 - |q_x|^2 + \frac{1}{2}(r_2 - r_1)\,|q|^2]\,dx\,dy\,.$$

We have the lemma [**121**].

LEMMA 4.11. *The inner product of the vector*

$$\mathcal{U} = \overline{\left(\begin{array}{c} \psi_2\hat{\psi}_2 \\ \psi_1\hat{\psi}_1 \end{array}\right)} + S\left(\begin{array}{c} \psi_2\hat{\psi}_2 \\ \psi_1\hat{\psi}_1 \end{array}\right),$$

where $\psi = (\psi_1, \psi_2)$ is an eigenfunction solving the Lax pair (3.26, 3.27), and $\hat{\psi} = (\hat{\psi}_1, \hat{\psi}_2)$ is an eigenfunction solving the corresponding congruent Lax pair (3.33, 3.34), and $S = \left(\begin{array}{cc} 0 & 1 \\ 1 & 0 \end{array}\right)$, with the vector field $J\nabla H$ given by the right hand side of (4.26) vanishes,

$$\langle \mathcal{U}, J\nabla H\rangle = 0,$$

where $J = \left(\begin{array}{cc} 0 & 1 \\ -1 & 0 \end{array}\right)$.

If we only consider even functions, i.e., q and $u = r_2 - r_1$ are even functions in both x and y, then we can split $\mathcal{U}$ into its even and odd parts,

$$\mathcal{U} = \mathcal{U}^{(e,x)}_{(e,y)} + \mathcal{U}^{(e,x)}_{(o,y)} + \mathcal{U}^{(o,x)}_{(e,y)} + \mathcal{U}^{(o,x)}_{(o,y)},$$

where

$$\mathcal{U}_{(e,y)}^{(e,x)} = \frac{1}{4}\Big[\mathcal{U}(x,y) + \mathcal{U}(-x,y) + \mathcal{U}(x,-y) + \mathcal{U}(-x,-y)\Big],$$

$$\mathcal{U}_{(o,y)}^{(e,x)} = \frac{1}{4}\Big[\mathcal{U}(x,y) + \mathcal{U}(-x,y) - \mathcal{U}(x,-y) - \mathcal{U}(-x,-y)\Big],$$

$$\mathcal{U}_{(e,y)}^{(o,x)} = \frac{1}{4}\Big[\mathcal{U}(x,y) - \mathcal{U}(-x,y) + \mathcal{U}(x,-y) - \mathcal{U}(-x,-y)\Big],$$

$$\mathcal{U}_{(o,y)}^{(o,x)} = \frac{1}{4}\Big[\mathcal{U}(x,y) - \mathcal{U}(-x,y) - \mathcal{U}(x,-y) + \mathcal{U}(-x,-y)\Big].$$

Then we have the lemma [121].

LEMMA 4.12. *When q and $u = r_2 - r_1$ are even functions in both x and y, we have*

$$\langle \mathcal{U}_{(e,y)}^{(e,x)} , J\nabla H \rangle = 0.$$

4.3.1. Melnikov Integrals.
Consider the perturbed DSII equation,

$$(4.28) \qquad \begin{cases} i\partial_t q = [\partial_x^2 - \partial_y^2]q + [2(|q|^2 - \omega^2) + u_y]q + \epsilon i f, \\ [\partial_x^2 + \partial_y^2]u = -4\partial_y |q|^2, \end{cases}$$

where q and u are respectively complex-valued and real-valued functions of three variables (t,x,y), and $G = (f,\overline{f})$ are the perturbation terms which can depend on q and $\overline{q}$ and their derivatives and t, x and y. The Melnikov integral is given by [121],

$$M = \int_{-\infty}^{\infty} \langle \mathcal{U}, G \rangle \, dt$$

$$(4.29) \qquad = 2\int_{-\infty}^{\infty}\int_0^{2\pi}\int_0^{2\pi} R_e \left\{ (\psi_2 \hat{\psi}_2)f + (\psi_1 \hat{\psi}_1)\overline{f} \right\} \, dx\, dy\, dt,$$

where the integrand is evaluated on an unperturbed homoclinic orbit in certain center-unstable (= center-stable) manifold, and such orbit can be obtained through the Bäcklund-Darboux transformations given in Theorem 3.6. A concrete example is given in section 3.3.1. When we only consider even functions, i.e., q and u are even functions in both x and y, the corresponding Melnikov function is given by [121],

$$M^{(e)} = \int_{-\infty}^{\infty} \langle \mathcal{U}_{(e,y)}^{(e,x)}, \vec{G} \rangle \, dt$$

$$(4.30) \qquad = \int_{-\infty}^{\infty} \langle \mathcal{U}, \vec{G} \rangle \, dt,$$

which is the same as expression (4.29).

4.3.2. An Example . We continue from the example in section 3.3.1. We generate the following eigenfunctions corresponding to the potential Q given in (3.55) through the iterated Bäcklund-Darboux transformations,

$$(4.31) \qquad \Psi^\pm \;=\; \Gamma^{(I)}\Gamma\psi^\pm, \qquad \text{at } \lambda = \lambda_0 = \sqrt{\eta^2 - \frac{1}{4}\kappa_1^2}\,,$$

$$(4.32) \qquad \Phi_\pm \;=\; \Gamma^{(I)}\Gamma\phi_\pm, \qquad \text{at } \lambda = 0\,,$$

where

$$\Gamma = \begin{bmatrix} \Lambda + a & b \\[2mm] \bar{b} & \Lambda - a \end{bmatrix}, \qquad \Gamma^{(I)} = \begin{bmatrix} \Lambda + a^{(I)} & b^{(I)} \\[2mm] \overline{b^{(I)}} & \Lambda - \overline{a^{(I)}} \end{bmatrix},$$

where $\Lambda = \alpha\partial_y - \lambda$ for general λ.

LEMMA 4.13 (see [**121**]). *The eigenfunctions $\Psi^\pm$ and $\Phi_\pm$ defined in (4.31) and (4.32) have the representations,*

$$(4.33) \qquad \Psi^\pm \;=\; \frac{\pm 2\lambda_0 W(\psi^+,\psi^-)}{\Delta} \begin{bmatrix} (-\lambda_0 + a^{(I)})\overline{\psi}_2 - b^{(I)}\overline{\psi}_1 \\[2mm] \overline{b^{(I)}}\,\overline{\psi}_2 + (\lambda_0 + \overline{a^{(I)}})\overline{\psi}_1 \end{bmatrix},$$

$$(4.34) \qquad \Phi_\pm \;=\; \frac{\mp i\alpha\kappa_2}{\Delta^{(I)}} \begin{bmatrix} \Xi_1 \\[2mm] \Xi_2 \end{bmatrix},$$

where

$$W(\psi^+,\psi^-) = \begin{vmatrix} \psi_1^+ & \psi_1^- \\[2mm] \psi_2^+ & \psi_2^- \end{vmatrix} = -i\kappa_1 c_0^+ c_0^- q_c \exp\{i2r_2 t\}\,,$$

$$\Delta = -\left[|\psi_1|^2 + |\psi_2|^2\right],$$

$$\psi = \psi^+ + \psi^-\,,$$

$$\Delta^{(I)} = -\left[|\varphi_1|^2 + |\varphi_2|^2\right],$$

$$\varphi = \varphi^+ + \varphi^-\,,$$

$$\varphi^\pm = \Gamma\phi_\pm \text{ at } \lambda = 0\,,$$

$$\Xi_1 = \overline{\varphi}_1(\varphi_1^+\varphi_1^-) + \overline{\varphi_2^+}(\varphi_1^+\varphi_2^-) + \overline{\varphi_2^-}(\varphi_1^-\varphi_2^+)\,,$$

$$\Xi_2 = \overline{\varphi}_2(\varphi_2^+\varphi_2^-) + \overline{\varphi_1^+}(\varphi_1^-\varphi_2^+) + \overline{\varphi_1^-}(\varphi_1^+\varphi_2^-)\,.$$

If we take r_2 to be real (in the Melnikov vectors, r_2 appears in the form $r_2 - r_1 = 2(|q_c|^2 - \omega^2)$), then

$$(4.35) \qquad \Psi^\pm \to 0, \qquad \Phi_\pm \to 0, \quad \text{as } t \to \pm\infty\,.$$

Next we generate eigenfunctions solving the corresponding congruent Lax pair (3.33, 3.34) with the potential Q, through the iterated Bäcklund-Darboux transformations and the symmetry transformation (3.32) [**121**].

LEMMA 4.14. *Under the replacements*

$$\alpha \longrightarrow -\alpha \;\; (\vartheta_2 \longrightarrow \pi - \vartheta_2), \quad \hat{\vartheta} \longrightarrow \hat{\vartheta} + \pi - 2\vartheta_2, \quad \hat{\rho} \longrightarrow -\hat{\rho}\,,$$

the coefficients in the iterated Bäcklund-Darboux transformations are transformed as follows,

$$a^{(I)} \longrightarrow \overline{a^{(I)}}, \quad b^{(I)} \longrightarrow b^{(I)},$$

$$\left(c^{(I)} = \overline{b^{(I)}}\right) \longrightarrow \left(c^{(I)} = \overline{b^{(I)}}\right), \quad \left(d^{(I)} = -\overline{a^{(I)}}\right) \longrightarrow \left(\overline{d^{(I)}} = -a^{(I)}\right).$$

LEMMA 4.15 (see [**121**]). *Under the replacements*

$$\alpha \mapsto -\alpha \quad (\vartheta_2 \longrightarrow \pi - \vartheta_2), \quad r_1 \mapsto -r_2,$$

(4.36)

$$r_2 \mapsto -r_1, \quad \hat{\vartheta} \longrightarrow \hat{\vartheta} + \pi - 2\vartheta_2, \quad \hat{\rho} \longrightarrow -\hat{\rho},$$

the potentials are transformed as follows,

$$\begin{aligned}
Q &\longrightarrow Q, \\
(R = \overline{Q}) &\longrightarrow (R = \overline{Q}), \\
R_1 &\longrightarrow -R_2, \\
R_2 &\longrightarrow -R_1.
\end{aligned}$$

The eigenfunctions $\Psi^\pm$ and $\Phi_\pm$ given in (4.33) and (4.34) depend on the variables in the replacement (4.36):

$$\begin{aligned}
\Psi^\pm &= \Psi^\pm(\alpha, r_1, r_2, \hat{\vartheta}, \hat{\rho}), \\
\Phi_\pm &= \Phi_\pm(\alpha, r_1, r_2, \hat{\vartheta}, \hat{\rho}).
\end{aligned}$$

Under replacement (4.36), $\Psi^\pm$ and $\Phi_\pm$ are transformed into

(4.37) $$\widehat{\Psi}^\pm = \Psi^\pm(-\alpha, -r_2, -r_1, \hat{\vartheta} + \pi - 2\vartheta_2, -\hat{\rho}),$$

(4.38) $$\widehat{\Phi}_\pm = \Phi_\pm(-\alpha, -r_2, -r_1, \hat{\vartheta} + \pi - 2\vartheta_2, -\hat{\rho}).$$

COROLLARY 4.16 (see [**121**]). $\widehat{\Psi}^\pm$ *and* $\widehat{\Phi}_\pm$ *solve the congruent Lax pair (3.33, 3.34) at* $(Q, \overline{Q}, R_1, R_2; \lambda_0)$ *and* $(Q, \overline{Q}, R_1, R_2; 0)$, *respectively.*

Notice that as a function of η, ξ_1^0 has two (plus and minus) branches. In order to construct Melnikov vectors, we need to study the effect of the replacement $\xi_1^0 \longrightarrow -\xi_1^0$.

LEMMA 4.17 (see [**121**]). *Under the replacements*

(4.39) $$\xi_1^0 \longrightarrow -\xi_1^0 \quad (\vartheta_2 \longrightarrow -\vartheta_2), \quad \hat{\vartheta} \longrightarrow \hat{\vartheta} + \pi - 2\vartheta_2, \quad \hat{\rho} \longrightarrow -\hat{\rho},$$

the coefficients in the iterated Bäcklund-Darboux transformations are invariant,

$$a^{(I)} \mapsto a^{(I)}, \quad b^{(I)} \mapsto b^{(I)},$$

$$(c^{(I)} = \overline{b^{(I)}}) \mapsto (c^{(I)} = \overline{b^{(I)}}), \quad (d^{(I)} = -\overline{a^{(I)}}) \mapsto (d^{(I)} = -\overline{a^{(I)}});$$

thus the potentials are also invariant,

$$Q \longrightarrow Q, \quad (R = \overline{Q}) \longrightarrow (R = \overline{Q}),$$

$$R_1 \longrightarrow R_1, \quad R_2 \longrightarrow R_2.$$

The eigenfunction $\Phi_\pm$ given in (4.34) depends on the variables in the replacement (4.39):

$$\Phi_\pm = \Phi_\pm(\xi_1^0, \hat{\vartheta}, \hat{\rho}) \,.$$

Under the replacement (4.39), $\Phi_\pm$ is transformed into

$$(4.40) \qquad \widetilde{\Phi}_\pm = \Phi_\pm(-\xi_1^0, \hat{\vartheta} + \pi - 2\vartheta_2, -\hat{\rho}) \,.$$

COROLLARY 4.18. $\widetilde{\Phi}_\pm$ *solves the Lax pair (3.26,3.27) at* $(Q, \overline{Q}, R_1, R_2\,;\,0)$.

In the construction of the Melnikov vectors, we need to replace $\Phi_\pm$ by $\widetilde{\Phi}_\pm$ to guarantee the periodicity in x of period $L_1 = \frac{2\pi}{\kappa_1}$.

The Melnikov vectors for the Davey-Stewartson II equations are given by,

$$(4.41) \qquad \mathcal{U}^\pm = \begin{pmatrix} \Psi_2^\pm \widehat{\Psi}_2^\pm \\ \Psi_1^\pm \widehat{\Psi}_1^\pm \end{pmatrix}^- + S \begin{pmatrix} \Psi_2^\pm \widehat{\Psi}_2^\pm \\ \Psi_1^\pm \widehat{\Psi}_1^\pm \end{pmatrix},$$

$$(4.42) \qquad \mathcal{U}_\pm = \begin{pmatrix} \widetilde{\Phi}_\pm^{(2)} \widehat{\Phi}_\pm^{(2)} \\ \widetilde{\Phi}_\pm^{(1)} \widehat{\Phi}_\pm^{(1)} \end{pmatrix}^- + S \begin{pmatrix} \widetilde{\Phi}_\pm^{(2)} \widehat{\Phi}_\pm^{(2)} \\ \widetilde{\Phi}_\pm^{(1)} \widehat{\Phi}_\pm^{(1)} \end{pmatrix},$$

where $S = \begin{pmatrix} 0 & 1 \\ 1 & 0 \end{pmatrix}$. The corresponding Melnikov functions (4.29) are given by,

$$M^\pm = \int_{-\infty}^{\infty} \langle \mathcal{U}^\pm, \vec{G} \rangle \, dt$$

$$= 2 \int_{-\infty}^{\infty} \int_0^{2\pi} \int_0^{2\pi} R_e \Big\{ [\Psi_2^\pm \widehat{\Psi}_2^\pm] f(Q, \overline{Q})$$

$$(4.43) \qquad \qquad + [\Psi_1^\pm \widehat{\Psi}_1^\pm] \overline{f}(Q, \overline{Q}) \Big\} \, dx \, dy \, dt \,,$$

$$M_\pm = \int_{-\infty}^{\infty} \langle \mathcal{U}_\pm, \vec{G} \rangle \, dt$$

$$= 2 \int_{-\infty}^{\infty} \int_0^{2\pi} \int_0^{2\pi} R_e \Big\{ [\widetilde{\Phi}_\pm^{(2)} \widehat{\Phi}_\pm^{(2)}] f(Q, \overline{Q})$$

$$(4.44) \qquad \qquad + [\widetilde{\Phi}_\pm^{(1)} \widehat{\Phi}_\pm^{(1)}] \overline{f}(Q, \overline{Q}) \Big\} \, dx \, dy \, dt \,,$$

where Q is given in (3.55), $\Psi^\pm$ is given in (4.33), $\widetilde{\Phi}_\pm$ is given in (4.34) and (4.40), $\widehat{\Psi}^\pm$ is given in (4.33) and (4.37), and $\widehat{\Phi}_\pm$ is given in (4.34) and (4.38). As given in (4.30), the above formulas also apply when we consider even function Q in both x and y.

CHAPTER 5

Invariant Manifolds

Invariant manifolds have attracted intensive studies which led to two main approaches: Hadamard's method [78] [64] and Perron's method [176] [42]. For example, for a partial differential equation of the form

$$\partial_t u = Lu + N(u) \,,$$

where L is a linear operator and $N(u)$ is the nonlinear term, if the following two ingredients

    (1) the gaps separating the unstable, center, and stable spectra of L are large enough,

    (2) the nonlinear term $N(u)$ is Lipschitzian in u with small Lipschitz constant,

are available, then establishing the existence of unstable, center, and stable manifolds is rather straightforward. Building invariant manifolds when any of the above conditions fails, is a very challenging and interesting problem [124].

There has been a vast literature on invariant manifolds. A good starting point of reading can be from the references [103] [64] [42]. Depending upon the emphasis on the specific problem, one may establish invariant manifolds for a specific flow, or investigate the persistence of existing invariant manifolds under perturbations to the flow.

In specific applications, most of the problems deal with manifolds with boundaries. In this context, the relevant concepts are overflowing invariant, inflowing invariant, and locally invariant submanifolds, defined in the Chapter on General Setup and Concepts.

5.1. Nonlinear Schrödinger Equation Under Regular Perturbations

Persistence of invariant manifolds depends upon the nature of the perturbation. Under the so-called regular perturbations, i.e., the perturbed evolution operator is C^1 close to the unperturbed one, for any fixed time; invariant manifolds persist "nicely". Under other singular perturbations, this may not be the case.

Consider the regularly perturbed nonlinear Schrödinger (NLS) equation [136] [140],

$$(5.1) \qquad iq_t = q_{xx} + 2[|q|^2 - \omega^2]q + i\epsilon[\hat{\partial}_x^2 q - \alpha q + \beta] \,,$$

where $q = q(t,x)$ is a complex-valued function of the two real variables t and x, t represents time, and x represents space. $q(t,x)$ is subject to periodic boundary condition of period 2π, and even constraint, i.e.,

$$q(t, x + 2\pi) = q(t, x) \,, \quad q(t, -x) = q(t, x) \,.$$

ω is a positive constant, $\alpha > 0$ and $\beta > 0$ are constants, $\hat{\partial}_x^2$ is a bounded Fourier multiplier,

$$\hat{\partial}_x^2 q = -\sum_{k=1}^{N} k^2 \xi_k \tilde{q}_k \cos kx ,$$

$\xi_k = 1$ when $k \leq N$, $\xi_k = 8k^{-2}$ when $k > N$, for some fixed large N, and $\epsilon > 0$ is the perturbation parameter.

One can prove the following theorems [**136**] [**140**].

THEOREM 5.1 (Persistence Theorem). *For any integers k and n ($1 \leq k, n < \infty$), there exist a positive constant ϵ_0 and a neighborhood $\mathcal{U}$ of the circle $S_\omega = \{q \mid \partial_x q = 0, \ |q| = \omega, \ 1/2 < \omega < 1\}$ in the Sobolev space H^k, such that inside $\mathcal{U}$, for any $\epsilon \in (-\epsilon_0, \epsilon_0)$, there exist C^n locally invariant submanifolds W_ϵ^{cu} and W_ϵ^{cs} of codimension 1, and W_ϵ^c ($= W_\epsilon^{cu} \cap W_\epsilon^{cs}$) of codimension 2 under the evolution operator F_ϵ^t given by (5.1). When $\epsilon = 0$, W_0^{cu}, W_0^{cs}, and W_0^c are tangent to the center-unstable, center-stable, and center subspaces of the circle of fixed points S_ω, respectively. W_ϵ^{cu}, W_ϵ^{cs}, and W_ϵ^c are C^n smooth in ϵ for $\epsilon \in (-\epsilon_0, \epsilon_0)$.*

W_ϵ^{cu}, W_ϵ^{cs}, and W_ϵ^c are called persistent center-unstable, center-stable, and center submanifolds near S_ω under the evolution operator F_ϵ^t given by (5.1).

THEOREM 5.2 (Fiber Theorem). *Inside the persistent center-unstable subman-ifold W_ϵ^{cu} near S_ω, there exists a family of 1-dimensional C^n smooth submanifolds (curves) $\{\mathcal{F}^{(u,\epsilon)}(q) : q \in W_\epsilon^c\}$, called unstable fibers:*

- W_ϵ^{cu} *can be represented as a union of these fibers,*

$$W_\epsilon^{cu} = \bigcup_{q \in W_\epsilon^c} \mathcal{F}^{(u,\epsilon)}(q).$$

- $\mathcal{F}^{(u,\epsilon)}(q)$ *depends C^{n-1} smoothly on both ϵ and q for $\epsilon \in (-\epsilon_0, \epsilon_0)$ and $q \in W_\epsilon^c$, in the sense that $\mathcal{W}$ defined by*

$$\mathcal{W} = \left\{ (q_1, q, \epsilon) \ \middle| \ q_1 \in \mathcal{F}^{(u,\epsilon)}(q), \ q \in W_\epsilon^c, \ \epsilon \in (-\epsilon_0, \epsilon_0) \right\}$$

is a C^{n-1} smooth submanifold of $H^k \times H^k \times (-\epsilon_0, \epsilon_0)$.
- *Each fiber $\mathcal{F}^{(u,\epsilon)}(q)$ intersects W_ϵ^c transversally at q, two fibers $\mathcal{F}^{(u,\epsilon)}(q_1)$ and $\mathcal{F}^{(u,\epsilon)}(q_2)$ are either disjoint or identical.*
- *The family of unstable fibers $\{\mathcal{F}^{(u,\epsilon)}(q) : q \in W_\epsilon^c\}$ is negatively invariant, in the sense that the family of fibers commutes with the evolution operator F_ϵ^t in the following way:*

$$F_\epsilon^t(\mathcal{F}^{(u,\epsilon)}(q)) \subset \mathcal{F}^{(u,\epsilon)}(F_\epsilon^t(q))$$

for all $q \in W_\epsilon^c$ and all $t \leq 0$ such that $\bigcup_{\tau \in [t,0]} F_\epsilon^\tau(q) \subset W_\epsilon^c$.
- *There are positive constants κ and C which are independent of ϵ such that if $q \in W_\epsilon^c$ and $q_1 \in \mathcal{F}^{(u,\epsilon)}(q)$, then*

$$\left\| F_\epsilon^t(q_1) - F_\epsilon^t(q) \right\| \leq C e^{\kappa t} \left\| q_1 - q \right\|,$$

for all $t \leq 0$ such that $\bigcup_{\tau \in [t,0]} F_\epsilon^\tau(q) \subset W_\epsilon^c$, where $\| \ \|$ denotes H^k norm of periodic functions of period 2π.

- *For any $q, p \in W_\epsilon^c$, $q \neq p$, any $q_1 \in \mathcal{F}^{(u,\epsilon)}(q)$ and any $p_1 \in \mathcal{F}^{(u,\epsilon)}(p)$; if*

$$F_\epsilon^t(q), F_\epsilon^t(p) \in W_\epsilon^c, \quad \forall t \in (-\infty, 0],$$

and

$$\|F_\epsilon^t(p_1) - F_\epsilon^t(q)\| \to 0, \quad as \ t \to -\infty;$$

then

$$\left\{ \frac{\|F_\epsilon^t(q_1) - F_\epsilon^t(q)\|}{\|F_\epsilon^t(p_1) - F_\epsilon^t(q)\|} \right\} \bigg/ e^{\frac{1}{2}\kappa t} \to 0, \quad as \ t \to -\infty.$$

Similarly for W_ϵ^{cs}.

When $\epsilon = 0$, certain low-dimensional invariant submanifolds of the invariant manifolds, have explicit representations through Darboux transformations. Specifically, the periodic orbit (3.9) where $1/2 < c < 1$, has two-dimensional stable and unstable manifolds given by (3.14). Unstable and stable fibers with bases along the periodic orbit also have expressions given by (3.14).

5.2. Nonlinear Schrödinger Equation Under Singular Perturbations

Consider the singularly perturbed nonlinear Schrödinger equation [124],

$$(5.2) \qquad iq_t = q_{xx} + 2[|q|^2 - \omega^2]q + i\epsilon[q_{xx} - \alpha q + \beta] \,,$$

where $q = q(t, x)$ is a complex-valued function of the two real variables t and x, t represents time, and x represents space. $q(t, x)$ is subject to periodic boundary condition of period 2π, and even constraint, i.e.,

$$q(t, x + 2\pi) = q(t, x) \,, \quad q(t, -x) = q(t, x) \,.$$

$\omega \in (1/2, 1)$ is a positive constant, $\alpha > 0$ and $\beta > 0$ are constants, and $\epsilon > 0$ is the perturbation parameter.

Here the perturbation term $\epsilon\partial_x^2$ generates the semigroup $e^{\epsilon t \partial_x^2}$, the regularity of the invariant mainfolds with respect to ϵ at $\epsilon = 0$ will be closely related to the regularity of the semigroup $e^{\epsilon t \partial_x^2}$ with respect to ϵ at $\epsilon = 0$. Also the singular perturbation term $\epsilon\partial_x^2 q$ breaks the spectral gap separating the center spectrum and the stable spectrum. Therefore, standard invariant manifold results do not apply. Invariant manifolds do not persist "nicely". Nevertheless, certain persistence results do hold.

One can prove the following unstable fiber theorem and center-stable manifold theorem [124].

THEOREM 5.3 (Unstable Fiber Theorem). *Let $\mathcal{A}$ be the annulus: $\mathcal{A} = \{q \mid \partial_x q = 0, \ 1/2 < |q| < 1\}$, for any $p \in \mathcal{A}$, there is an unstable fiber $\mathcal{F}_p^+$ which is a curve. $\mathcal{F}_p^+$ has the following properties:*

(1) *$\mathcal{F}_p^+$ is a C^1 smooth in H^k-norm, $k \geq 1$.*
(2) *$\mathcal{F}_p^+$ is also C^1 smooth in ϵ, α, β, ω, and p in H^k-norm, $k \geq 1$, $\epsilon \in [0, \epsilon_0)$ for some $\epsilon_0 > 0$.*
(3) *$\mathcal{F}_p^+$ has the exponential decay property: $\forall p_1 \in \mathcal{F}_p^+$,*

$$\frac{\|F^t p_1 - F^t p\|_k}{\|p_1 - p\|_k} \leq C e^{\mu t}, \quad \forall t \leq 0,$$

where F^t is the evolution operator, $\mu > 0$.

(4) $\{\mathcal{F}_p^+\}_{p\in\mathcal{A}}$ *forms an invariant family of unstable fibers,*

$$F^t \mathcal{F}_p^+ \subset \mathcal{F}_{F^t p}^+ \ , \quad \forall t \in [-T, 0],$$

and $\forall T > 0$ *(T can be $+\infty$), such that $F^t p \in \mathcal{A}$, $\forall t \in [-T, 0]$.*

THEOREM 5.4 (Center-Stable Manifold Theorem). *There exists a C^1 smooth codimension 1 locally invariant center-stable manifold W_k^{cs} in a neighborhood of the annulus $\mathcal{A}$ (Theorem 5.3) in H^k for any $k \geq 1$.*

   (1) *At points in the subset W_{k+4}^{cs} of W_k^{cs}, W_k^{cs} is C^1 smooth in ϵ in H^k-norm for $\epsilon \in [0, \epsilon_0)$ and some $\epsilon_0 > 0$.*
   (2) *W_k^{cs} is C^1 smooth in (α, β, ω).*

REMARK 5.5. C^1 regularity in ϵ is crucial in locating a homoclinic orbit. As can be seen later, one has detailed information on certain unperturbed (i.e. $\epsilon = 0$) homoclinic orbit, which will be used in tracking candidates for a perturbed homoclinic orbit. In particular, Melnikov measurement will be needed. Melnikov measurement measures zeros of $\mathcal{O}(\epsilon)$ signed distances, thus, the perturbed orbit needs to be $\mathcal{O}(\epsilon)$ close to the unperturbed orbit in order to perform Melnikov measurement.

5.3. Proof of the Unstable Fiber Theorem 5.3

Here we give the proof of the unstable fiber theorem 5.3, proofs of other fiber theorems in this chapter are easier.

5.3.1. The Setup of Equations. First, write q as

$$(5.3) \qquad q(t, x) = [\rho(t) + f(t, x)]e^{i\theta(t)},$$

where f has zero spatial mean. We use the notation $\langle \cdot \rangle$ to denote spatial mean,

$$(5.4) \qquad \langle q \rangle = \frac{1}{2\pi} \int_0^{2\pi} q\, dx.$$

Since the L^2-norm is an action variable when $\epsilon = 0$, it is more convenient to replace ρ by:

$$(5.5) \qquad I = \langle |q|^2 \rangle = \rho^2 + \langle |f|^2 \rangle.$$

The final pick is

$$(5.6) \qquad J = I - \omega^2.$$

In terms of the new variables (J, θ, f), Equation (5.2) can be rewritten as

$$(5.7) \qquad \dot{J} = \epsilon\left[-2\alpha(J + \omega^2) + 2\beta\sqrt{J + \omega^2}\cos\theta\right] + \epsilon\mathcal{R}_2^J,$$

$$(5.8) \qquad \dot{\theta} = -2J - \epsilon\beta\frac{\sin\theta}{\sqrt{J + \omega^2}} + \mathcal{R}_2^\theta,$$

$$(5.9) \qquad f_t = L_\epsilon f + V_\epsilon f - i\mathcal{N}_2 - i\mathcal{N}_3,$$

where

$$(5.10) \qquad L_\epsilon f = -i f_{xx} + \epsilon(-\alpha f + f_{xx}) - i2\omega^2(f + \bar f),$$

$$(5.11) \qquad V_\epsilon f = -i2J(f + \bar f) + i\epsilon\beta f \frac{\sin\theta}{\sqrt{J + \omega^2}},$$

$$(5.12) \qquad \mathcal{R}_2^J = -2\langle |f_x|^2 \rangle + 2\beta\cos\theta \left[\sqrt{J + \omega^2 - \langle |f|^2 \rangle} - \sqrt{J + \omega^2} \right],$$

$$\mathcal{R}_2^\theta = -\langle (f + \bar f)^2 \rangle - \frac{1}{\rho}\langle |f|^2 (f + \bar f) \rangle$$

$$(5.13) \qquad\qquad - \epsilon\beta\sin\theta \left[\frac{1}{\sqrt{J + \omega^2 - \langle |f|^2 \rangle}} - \frac{1}{\sqrt{J + \omega^2}} \right],$$

$$(5.14) \qquad \mathcal{N}_2 = 2\rho[2(|f|^2 - \langle |f|^2 \rangle) + (f^2 - \langle f^2 \rangle)],$$

$$\mathcal{N}_3 = -\langle f^2 + \bar f^2 + 6|f|^2 \rangle f + 2(|f|^2 f - \langle |f|^2 f \rangle)$$

$$- \frac{1}{\rho}\langle |f|^2 (f + \bar f) \rangle f - 2\langle |f|^2 \rangle \bar f$$

$$(5.15) \qquad\qquad - \epsilon\beta\sin\theta \left[\frac{1}{\sqrt{J + \omega^2 - \langle |f|^2 \rangle}} - \frac{1}{\sqrt{J + \omega^2}} \right] f.$$

REMARK 5.6. The singular perturbation term "$\epsilon\partial_x^2 q$" can be seen at two locations, L_ϵ and $\mathcal{R}_2^J$ (5.10,5.12). The singular perturbation term $\langle |f_x|^2 \rangle$ in $\mathcal{R}_2^J$ does not create any difficulty. Since H^1 is a Banach algebra [6], this term is still of quadratic order, $\langle |f_x|^2 \rangle \sim \mathcal{O}(\|f\|_1^2)$.

LEMMA 5.7. *The nonlinear terms have the orders:*

$$|\mathcal{R}_2^J| \sim \mathcal{O}(\|f\|_s^2), \quad |\mathcal{R}_2^\theta| \sim \mathcal{O}(\|f\|_s^2),$$

$$\|\mathcal{N}_2\|_s \sim \mathcal{O}(\|f\|_s^2), \quad \|\mathcal{N}_3\|_s \sim \mathcal{O}(\|f\|_s^3), \quad (s \geq 1).$$

Proof: The proof is an easy direct verification. Q.E.D.

5.3.2. The Spectrum of L_ϵ. The spectrum of L_ϵ consists of only point spectrum. The eigenvalues of L_ϵ are:

$$(5.16) \qquad \mu_k^\pm = -\epsilon(\alpha + k^2) \pm k\sqrt{4\omega^2 - k^2}, \quad (k = 1, 2, \ldots);$$

where $\omega \in \left(\frac{1}{2}, 1\right)$, only $\mu_1^\pm$ are real, and $\mu_k^\pm$ are complex for $k > 1$.

The main difficulty introduced by the singular perturbation $\epsilon\partial_x^2 f$ is the breaking of the spectral gap condition. Figure 1 shows the distributions of the eigenvalues when $\epsilon = 0$ and $\epsilon \neq 0$. It clearly shows the breaking of the stable spectral gap condition. As a result, center and center-unstable manifolds do not necessarily persist. On the other hand, the unstable spectral gap condition is not broken. This gives the hope for the persistence of center-stable manifold. Another case of persistence can be described as follows: Notice that the plane Π,

$$(5.17) \qquad\qquad \Pi = \{q \mid \partial_x q = 0\},$$

is invariant under the flow (5.2). When $\epsilon = 0$, there is an unstable fibration with base points in a neighborhood of the circle S_ω of fixed points,

$$(5.18) \qquad\qquad S_\omega = \{q \in \Pi \mid |q| = \omega\},$$

in Π, as an invariant sub-fibration of the unstable Fenichel fibration with base points in the center manifold. When $\epsilon > 0$, the center manifold may not persist, but Π

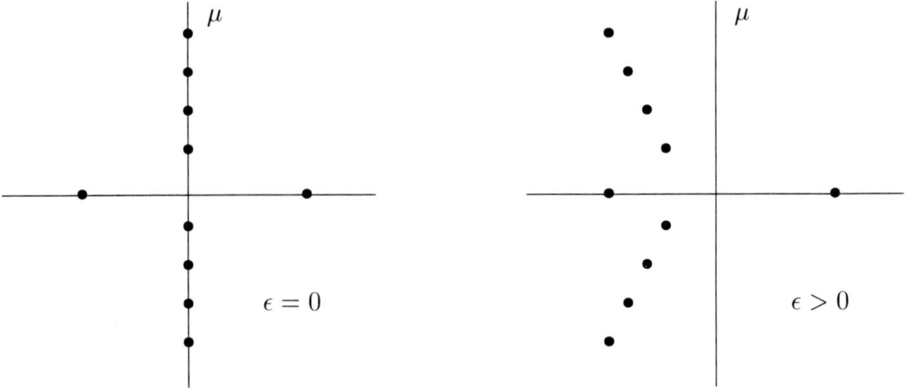

FIGURE 1. The point spectra of the linear operator L_ϵ.

persists, moreover, the unstable spectral gap condition is not broken, therefore, the unstable sub-fibration with base points in Π may persist. Since the semiflow generated by (5.2) is not a C^1 perturbation of that generated by the unperturbed NLS due to the singular perturbation $\epsilon\partial_x^2$, standard results on persistence can not be applied.

The eigenfunctions corresponding to the real eigenvalues are:

$$(5.19) \qquad e_1^\pm = e^{\pm i\vartheta_1}\cos x, \quad e^{\pm i\vartheta_1} = \frac{1 \mp i\sqrt{4\omega^2 - 1}}{2\omega}.$$

Notice that they are independent of ϵ. The eigenspaces corresponding to the complex conjugate pairs of eigenvalues are given by:

$$E_1 = \operatorname{span}_{\mathbb{C}}\{\cos x\}.$$

and have real dimension 2.

5.3.3. The Re-Setup of Equations. For the goal of this subsection, we need to single out the eigen-directions (5.19). Let

$$f = \sum_{\pm} \xi_1^\pm e_1^\pm + h,$$

where $\xi_1^\pm$ are real variables, and

$$\langle h \rangle = \langle h \cos x \rangle = 0.$$

In terms of the coordinates $(\xi_1^\pm, J, \theta, h)$, (5.7)-(5.9) can be rewritten as:

$$(5.20) \qquad \dot{\xi}_1^+ = \mu_1^+ \xi_1^+ + V_1^+ \xi_1^+ + \mathcal{N}_1^+,$$

$$(5.21) \qquad \dot{J} = \epsilon\left[-2\alpha(J + \omega^2) + 2\beta\sqrt{J + \omega^2}\cos\theta\right] + \epsilon\mathcal{R}_2^J,$$

$$(5.22) \qquad \dot{\theta} = -2J - \epsilon\beta\frac{\sin\theta}{\sqrt{J + \omega^2}} + \mathcal{R}_2^\theta,$$

$$(5.23) \qquad h_t = L_\epsilon h + V_\epsilon h + \tilde{\mathcal{N}},$$

$$(5.24) \qquad \dot{\xi}_1^- = \mu_1^- \xi_1^- + V_1^- \xi_1^- + \mathcal{N}_1^-,$$

where $\mu_1^{\pm}$ are given in (5.16), $\mathcal{N}_1^{\pm}$ and $\tilde{\mathcal{N}}$ are projections of $-i\mathcal{N}_2 - i\mathcal{N}_3$ to the corresponding directions, and

$$V_1^+ \xi_1^+ = 2c_1 J(\xi_1^+ + \xi_1^-) + \epsilon\beta \frac{\sin\theta}{\sqrt{J + \omega^2}}(c_1^+ \xi_1^+ - c_1^- \xi_1^-),$$

$$V_1^- \xi_1^- = -2c_1 J(\xi_1^+ + \xi_1^-) + \epsilon\beta \frac{\sin\theta}{\sqrt{J + \omega^2}}(c_1^- \xi_1^+ - c_1^+ \xi_1^-),$$

$$c_1 = \frac{1}{\sqrt{4\omega^2 - 1}}, \quad c_1^+ = \frac{2\omega^2 - 1}{\sqrt{4\omega^2 - 1}}, \quad c_1^- = \frac{2\omega^2}{\sqrt{4\omega^2 - 1}}.$$

5.3.4. A Modification.

DEFINITION 5.8. For any $\delta > 0$, we define the annular neighborhood of the circle S_ω (5.18) as

$$\mathcal{A}(\delta) = \{(J, \theta) \mid |J| < \delta\}.$$

To apply the Lyapunov-Perron's method, it is standard and necessary to modify the J equation so that $\mathcal{A}(4\delta)$ is overflowing invariant. Let $\eta \in C^\infty(R, R)$ be a "bump" function:

$$\eta = \begin{cases} 0, & \text{in } (-2, 2) \cup (-\infty, -6) \cup (6, \infty), \\ 1, & \text{in } (3, 5), \\ -1, & \text{in } (-5, -3), \end{cases}$$

and $|\eta'| \leq 2$, $|\eta''| \leq C$. We modify the J equation (5.21) as follows:

$$(5.25) \qquad \dot{J} = \epsilon b\eta(J/\delta) + \epsilon\left[-2\alpha(J + \omega^2) + 2\beta\sqrt{J + \omega^2}\cos\theta\right] + \epsilon\mathcal{R}_2^J,$$

where $b > 2(2\alpha\omega^2 + 2\beta\omega)$. Then $\mathcal{A}(4\delta)$ is overflowing invariant. There are two main points in adopting the bump function:

(1) One needs $\mathcal{A}(4\delta)$ to be overflowing invariant so that a Lyapunov-Perron type integral equation can be set up along orbits in $\mathcal{A}(4\delta)$ for $t \in (-\infty, 0)$.
(2) One needs the vector field inside $\mathcal{A}(2\delta)$ to be unchanged so that results for the modified system can be claimed for the original system in $\mathcal{A}(\delta)$.

REMARK 5.9. Due to the singular perturbation, the real part of $\mu_k^{\pm}$ approaches $-\infty$ as $k \to \infty$. Thus the h equation (5.23) can not be modified to give overflowing flow. This rules out the construction of unstable fibers with base points having general h coordinates.

5.3.5. Existence of Unstable Fibers.
For any $(J_0, \theta_0) \in \mathcal{A}(4\delta)$, let

$$(5.26) \qquad J = J_*(t), \quad \theta = \theta_*(t), \quad t \in (-\infty, 0],$$

be the backward orbit of the modified system (5.25) and (5.22) with the initial point (J_0, θ_0). If

$$(\xi_1^+(t), J_*(t) + \tilde{J}(t), \theta_*(t) + \tilde{\theta}(t), h(t), \xi_1^-(t))$$

is a solution of the modified full system, then one has

$$(5.27) \qquad \dot{\xi}_1^+ = \mu_1^+ \xi_1^+ + F_1^+,$$

$$(5.28) \qquad u_t = Au + F,$$

where

$$u = \begin{pmatrix} \tilde{J} \\ \tilde{\theta} \\ h \\ \xi_1^- \end{pmatrix}, \quad A = \begin{pmatrix} 0 & 0 & 0 & 0 \\ -2 & 0 & 0 & 0 \\ 0 & 0 & L_\epsilon & 0 \\ 0 & 0 & 0 & \mu_1^- \end{pmatrix}, \quad F = \begin{pmatrix} F_J \\ F_\theta \\ F_h \\ F_1^- \end{pmatrix},$$

$$F_1^+ = V_1^+ \xi_1^+ + \mathcal{N}_1^+,$$

$$F_J = \epsilon b \left[\eta(J/\delta) - \eta(J_*(t)/\delta) \right] + \epsilon \left[-2\alpha \tilde{J} + 2\beta \sqrt{J + \omega^2} \cos\theta \right.$$

$$\left. - 2\beta \sqrt{J_*(t) + \omega^2} \cos\theta_*(t) \right] + \epsilon \mathcal{R}_2^J,$$

$$F_\theta = -\epsilon\beta \frac{\sin\theta}{\sqrt{J + \omega^2}} + \epsilon\beta \frac{\sin\theta_*(t)}{\sqrt{J_*(t) + \omega^2}} + \mathcal{R}_2^\theta,$$

$$F_h = V_\epsilon h + \tilde{\mathcal{N}},$$

$$F_1^- = V_1^- \xi_1^- + \mathcal{N}_1^-,$$

$$J = J_*(t) + \tilde{J}, \quad \theta = \theta_*(t) + \tilde{\theta}.$$

System (5.27)-(5.28) can be written in the equivalent integral equation form:

$$(5.29) \qquad \xi_1^+(t) = \xi_1^+(t_0)e^{\mu_1^+(t-t_0)} + \int_{t_0}^t e^{\mu_1^+(t-\tau)} F_1^+(\tau)d\tau,$$

$$(5.30) \qquad u(t) = e^{A(t-t_0)}u(t_0) + \int_{t_0}^t e^{A(t-\tau)} F(\tau)d\tau.$$

By virtue of the gap between μ_1^+ and the real parts of the eigenvalues of A, one can introduce the following space: For $\sigma \in \left(\frac{\mu_1^+}{100}, \frac{\mu_1^+}{3} \right)$, and $n \geq 1$, let

$$G_{\sigma,n} = \left\{ g(t) = (\xi_1^+(t), u(t)) \, \middle| \, t \in (-\infty, 0], \; g(t) \text{ is continuous} \right.$$

$$\left. \text{in } t \text{ in } H^n \text{ norm }, \; \|g\|_{\sigma,n} = \sup_{t \leq 0} e^{-\sigma t}[|\xi_1^+(t)| + \|u(t)\|_n] < \infty \right\}.$$

$G_{\sigma,n}$ is a Banach space under the norm $\|\cdot\|_{\sigma,n}$. Let $\mathcal{B}_{\sigma,n}(r)$ denote the ball in $G_{\sigma,n}$ centered at the origin with radius r. Since A only has point spectrum, the spectral mapping theorem is valid. It is obvious that for $t \geq 0$,

$$\|e^{At}u\|_n \leq C(1+t)\|u\|_n,$$

for some constant C. Thus, if $g(t) \in \mathcal{B}_{\sigma,n}(r)$, $r < \infty$ is a solution of (5.29)-(5.30), by letting $t_0 \to -\infty$ in (5.30) and setting $t_0 = 0$ in (5.29), one has

$$(5.31) \qquad \xi_1^+(t) = \xi_1^+(0)e^{\mu_1^+ t} + \int_0^t e^{\mu_1^+(t-\tau)} F_1^+(\tau)d\tau,$$

$$(5.32). \qquad u(t) = \int_{-\infty}^t e^{A(t-\tau)} F(\tau)d\tau.$$

For $g(t) \in \mathcal{B}_{\sigma,n}(r)$, let $\Gamma(g)$ be the map defined by the right hand side of (5.31)-(5.32). Then a solution of (5.31)-(5.32) is a fixed point of Γ. For any $n \geq 1$ and $\epsilon < \delta^2$, and δ and r are small enough, F_1^+ and F are Lipschitz in g with small Lipschitz constants. Standard arguments of the Lyapunov-Perron's method readily imply the existence of a fixed point g_* of Γ in $\mathcal{B}_{\sigma,n}(r)$.

5.3.6. Regularity of the Unstable Fibers in ϵ. The difficulties lie in the investigation on the regularity of g_* with respect to $(\epsilon, \alpha, \beta, \omega, J_0, \theta_0, \xi_1^+(0))$. The most difficult one is the regularity with respect to ϵ due to the singular perturbation. Formally differentiating g_* in (5.31)-(5.32) with respect to ϵ, one gets

$$(5.33) \qquad \xi_{1,\epsilon}^+(t) = \int_0^t e^{\mu_1^+(t-\tau)}\left[\partial_u F_1^+ \cdot u_\epsilon + \partial_{\xi_1^+} F_1^+ \cdot \xi_{1,\epsilon}^+\right](\tau)d\tau + \mathcal{R}_1^+(t),$$

$$(5.34) \qquad u_\epsilon(t) = \int_{-\infty}^t e^{A(t-\tau)}\left[\partial_u F \cdot u_\epsilon + \partial_{\xi_1^+} F \cdot \xi_{1,\epsilon}^+\right](\tau)d\tau + \mathcal{R}(t),$$

where

$$(5.35) \qquad \begin{aligned} \mathcal{R}_1^+(t) &= \xi_1^+(0)\mu_{1,\epsilon}^+ t e^{\mu_1^+ t} + \int_0^t \mu_{1,\epsilon}^+(t-\tau)e^{\mu_1^+(t-\tau)}F_1^+(\tau)d\tau \\ &\quad + \int_0^t e^{\mu_1^+(t-\tau)}[\partial_\epsilon F_1^+ + \partial_{u_*} F_1^+ \cdot u_{*,\epsilon}](\tau)d\tau, \end{aligned}$$

$$(5.36) \qquad \begin{aligned} \mathcal{R}(t) &= \int_{-\infty}^t (t-\tau)A_\epsilon e^{A(t-\tau)}F(\tau)d\tau \\ &\quad + \int_{-\infty}^+ e^{A(t-\tau)}[\partial_\epsilon F + \partial_{u_*} F \cdot u_{*,\epsilon}](\tau)d\tau, \end{aligned}$$

$$(5.37) \qquad \mu_{1,\epsilon}^+ = -(\alpha + 1),$$

$$(5.38) \qquad A_\epsilon = \begin{pmatrix} 0 & 0 & 0 & 0 \\ 0 & 0 & 0 & 0 \\ 0 & 0 & -\alpha + \partial_x^2 & 0 \\ 0 & 0 & 0 & -(\alpha+1) \end{pmatrix},$$

$$(5.39) \qquad u_* = (J_*, \theta_*, 0, 0)^T,$$

where $T =$ transpose, and (J_*, θ_*) are given in (5.26). The troublesome terms are the ones containing A_ϵ or $u_{*,\epsilon}$ in (5.35)-(5.36).

$$(5.40) \qquad \|A_\epsilon F\|_n \leq C\|F\|_{n+2} \leq \tilde{c}(\|u\|_{n+2} + |\xi_1^+|),$$

where $\tilde{c}$ is small when $(\,\cdot\,)$ on the right hand side is small.

$$\partial_{J_*} F_J \cdot J_{*,\epsilon} = \frac{\epsilon}{\delta}b[\eta'(J/\delta) - \eta'(J_*/\delta)]J_{*,\epsilon}$$
$$+ \epsilon[\beta\frac{\cos\theta}{\sqrt{J+\omega^2}} - \beta\frac{\cos\theta_*}{\sqrt{J_*+\omega^2}}]J_{*,\epsilon}$$
$$+ \text{ easier terms.}$$

$$|\partial_{J_*} F_J \cdot J_{*,\epsilon}| \leq \frac{\epsilon}{\delta^2}b \sup_{0\leq\hat{\gamma}\leq 1} |\eta''([\hat{\gamma}J_* + (1-\hat{\gamma})J]/\delta)| \, |\tilde{J}| \, |J_{*,\epsilon}|$$
$$+ \epsilon\beta C(|\tilde{J}| + |\tilde{\theta}|)|J_{*,\epsilon}| + \text{ easier terms}$$
$$\leq C_1(|\tilde{J}| + |\tilde{\theta}|)|J_{*,\epsilon}| + \text{ easier terms,}$$

$$\sup_{t\leq 0} e^{-\sigma t}|\partial_{J_*} F_J \cdot J_{*,\epsilon}| \leq C_1 \sup_{t\leq 0}[e^{-(\sigma+\tilde{\nu})t}(|\tilde{J}| + |\tilde{\theta}|)] \sup_{t\leq 0}[e^{\tilde{\nu}t}|J_{*,\epsilon}|]$$
$$+ \text{ easier terms,}$$

where $\sup_{t\leq 0} e^{\tilde{\nu}t}|J_{*,\epsilon}|$ can be bounded when ϵ is sufficiently small for any fixed $\tilde{\nu} > 0$, through a routine estimate on Equations (5.25) and (5.22) for $(J_*(t), \theta_*(t))$. Other terms involving $u_{*,\epsilon}$ can be estimated similarly. Thus, the $\| \ \|_{\sigma,n}$ norm of terms involving $u_{*,\epsilon}$ has to be bounded by $\| \ \|_{\sigma+\tilde{\nu},n}$ norms. This leads to the standard rate condition for the regularity of invariant manifolds. That is, the regularity is controlled by the spectral gap. The $\| \ \|_{\sigma,n}$ norm of the term involving A_ϵ has to be bounded by $\| \ \|_{\sigma,n+2}$ norms. This is a new phenomenon caused by the singular perturbation. This problem is resolved by virtue of a special property of the fixed point g_* of Γ. Notice that if $\sigma_2 \geq \sigma_1$, $n_2 \geq n_1$, then $G_{\sigma_2,n_2} \subset G_{\sigma_1,n_1}$. Thus by the uniqueness of the fixed point, if g_* is the fixed point of Γ in G_{σ_2,n_2}, g_* is also the fixed point of Γ in G_{σ_1,n_1}. Since g_* exists in $G_{\sigma,n}$ for an fixed $n \geq 1$ and $\sigma \in (\frac{\mu_1^+}{100}, \frac{\mu_1^+}{3} - 10\tilde{\nu})$ where $\tilde{\nu}$ is small enough,

$$\|\mathcal{R}_1^+\|_{\sigma,n} \leq C_1 + C_2\|g_*\|_{\sigma+\tilde{\nu},n},$$
$$\|\mathcal{R}\|_{\sigma,n} \leq C_3\|g_*\|_{\sigma,n+2} + C_4\|g_*\|_{\sigma+\tilde{\nu},n} + C_5,$$

where C_j $(1 \leq j \leq 5)$ depend upon $\|g_*(0)\|_n$ and $\|g_*(0)\|_{n+2}$. Let

$$M = 2(\|\mathcal{R}_1^+\|_{\sigma,n} + \|\mathcal{R}\|_{\sigma,n}),$$

and Γ' denote the linear map defined by the right hand sides of (5.33) and (5.34). Since the terms $\partial_u F_1^+$, $\partial_{\xi_1^+} F_1^+$, $\partial_u F$, and $\partial_{\xi_1^+} F$ all have small $\| \ \|_n$ norms, Γ' is a contraction map on $\mathcal{B}(M) \subset L(\mathcal{R}, G_{\sigma,n})$, where $\mathcal{B}(M)$ is the ball of radius M. Thus Γ' has a unique fixed point $g_{*,\epsilon}$. Next one needs to show that $g_{*,\epsilon}$ is indeed the partial derivative of g_* with respect to ϵ. That is, one needs to show

$$(5.41) \qquad \lim_{\Delta\epsilon \to 0} \frac{\|g_*(\epsilon + \Delta\epsilon) - g_*(\epsilon) - g_{*,\epsilon}\Delta\epsilon\|_{\sigma,n}}{\Delta\epsilon} = 0.$$

This has to be accomplished directly from Equations (5.31)-(5.32), (5.33)-(5.34) satisfied by g_* and $g_{*,\epsilon}$. The most troublesome estimate is still the one involving A_ϵ. First, notice the fact that $e^{\epsilon\partial_x^2}$ is holomorphic in ϵ when $\epsilon > 0$, and not differentiable at $\epsilon = 0$. Then, notice that $g_* \in G_{\sigma,n}$ for any $n \geq 1$, thus, $e^{\epsilon\partial_x^2} g_*$ is differentiable, up to certain order m, in ϵ at $\epsilon = 0$ from the right, i.e.

$$(d^+/d\epsilon)^m e^{\epsilon\partial_x^2} g_*|_{\epsilon=0}$$

exists in H^n. Let

$$z(t, \Delta\epsilon) = e^{(\epsilon+\Delta\epsilon)t\partial_x^2} g_* - e^{\epsilon t\partial_x^2} g_* - (\Delta\epsilon)t\partial_x^2 e^{\epsilon t\partial_x^2} g_*$$
$$= e^{\epsilon t\partial_x^2} w(\Delta\epsilon),$$

where $t \geq 0$, $\Delta\epsilon > 0$, and

$$w(\Delta\epsilon) = e^{(\Delta\epsilon)t\partial_x^2} g_* - g_* - (\Delta\epsilon)t\partial_x^2 g_*.$$

Since $w(0) = 0$, by the Mean Value Theorem, one has

$$\|w(\Delta\epsilon)\|_n = \|w(\Delta\epsilon) - w(0)\|_n \leq \sup_{0 \leq \lambda \leq 1} \|\frac{dw}{d\Delta\epsilon}(\lambda\Delta\epsilon)\|_n |\Delta\epsilon|,$$

where at $\lambda = 0$, $\frac{d}{d\Delta\epsilon} = \frac{d^+}{d\Delta\epsilon}$, and

$$\frac{dw}{d\Delta\epsilon} = t[e^{(\Delta\epsilon)t\partial_x^2}\partial_x^2 g_* - \partial_x^2 g_*].$$

Since $\frac{dw}{d\Delta\epsilon}(0) = 0$, by the Mean Value Theorem again, one has

$$\|\frac{dw}{d\Delta\epsilon}(\lambda\Delta\epsilon)\|_n = \|\frac{dw}{d\Delta\epsilon}(\lambda\Delta\epsilon) - \frac{dw}{d\Delta\epsilon}(0)\|_n \leq \sup_{0\leq\lambda_1\leq 1}\|\frac{d^2w}{d\Delta\epsilon^2}(\lambda_1\lambda\Delta\epsilon)\|_n|\lambda||\Delta\epsilon| \, ,$$

where

$$\frac{d^2w}{d\Delta\epsilon^2} = t^2[e^{(\Delta\epsilon)t\partial_x^2}\partial_x^4 g_*] \, .$$

Therefore, one has the estimate

(5.42) $$\|z(t,\Delta\epsilon)\|_n \leq |\Delta\epsilon|^2 t^2 \|g_*\|_{n+4} \, .$$

This estimate is sufficient for handling the estimate involving A_ϵ. The estimate involving $u_{*,\epsilon}$ can be handled in a similar manner. For instance, let

$$\tilde{z}(t,\Delta\epsilon) = F(u_*(t,\epsilon + \Delta\epsilon)) - F(u_*(t,\epsilon)) - \Delta\epsilon\partial_{u_*}F \cdot u_{*,\epsilon} \, ,$$

then

$$\|\tilde{z}(t,\Delta\epsilon)\|_{\sigma,n} \leq |\Delta\epsilon|^2 \sup_{0\leq\lambda\leq 1}\|[u_{*,\epsilon}\cdot\partial_{u_*}^2 F\cdot u_{*,\epsilon} + \partial_{u_*}F\cdot u_{*,\epsilon\epsilon}](\lambda\Delta\epsilon)\|_{\sigma,n} \, .$$

From the expression of F (5.28), one has

$$\|u_{*,\epsilon}\cdot\partial_{u_*}^2 F\cdot u_{*,\epsilon} + \partial_{u_*}F\cdot u_{*,\epsilon\epsilon}\|_{\sigma,n}$$
$$\leq C_1\|g_*\|_{\sigma+2\tilde{\nu},n}[(\sup_{t\leq 0}e^{\tilde{\nu}t}|u_{*,\epsilon}|)^2 + \sup_{t\leq 0}e^{2\tilde{\nu}t}|u_{*,\epsilon\epsilon}|],$$

and the term $[\ ]$ on the right hand side can be easily shown to be bounded. In conclusion, let

$$h = g_*(\epsilon + \Delta\epsilon) - g_*(\epsilon) - g_{*,\epsilon}\Delta\epsilon,$$

one has the estimate

$$\|h\|_{\sigma,n} \leq \tilde{\kappa}\|h\|_{\sigma,n} + |\Delta\epsilon|^2\tilde{C}(\|g_*\|_{\sigma,n+4}; \|g_*\|_{\sigma+2\tilde{\nu},n}),$$

where $\tilde{\kappa}$ is small, thus

$$\|h\|_{\sigma,n} \leq 2|\Delta\epsilon|^2\tilde{C}(\|g_*\|_{\sigma,n+4}; \|g_*\|_{\sigma+2\tilde{\nu},n}).$$

This implies that

$$\lim_{\Delta\epsilon\to 0}\frac{\|h\|_{\sigma,n}}{|\Delta\epsilon|} = 0,$$

which is (5.41).

Let $g_*(t) = (\xi_1^+(t), u(t))$. First, let me comment on $\frac{\partial u}{\partial\xi_1^+(0)}\big|_{\xi_1^+(0)=0,\epsilon=0} = 0$. From (5.32), one has

$$\|\frac{\partial u}{\partial\xi_1^+(0)}\|_{\sigma,n} \leq \kappa_1\|\frac{\partial g_*}{\partial\xi_1^+(0)}\|_{\sigma,n},$$

where by letting $\xi_1^+(0) \to 0$ and $\epsilon \to 0^+$, $\kappa_1 \to 0$. Thus

$$\frac{\partial u}{\partial\xi_1^+(0)}\bigg|_{\xi_1^+(0)=0,\epsilon=0} = 0.$$

I shall also comment on "exponential decay" property. Since $\|g_*\|_{\frac{\mu_1^+}{3},n} \leq r$,

$$\|g_*(t)\|_n \leq re^{\frac{\mu_1^+}{3}t}, \quad \forall t \leq 0.$$

DEFINITION 5.10. Let $g_*(t) = (\xi_1^+(t), u(t))$, where

$$u(0) = \int_{-\infty}^{0} e^{A(t-\tau)} F(\tau) d\tau$$

depends upon $\xi_1^+(0)$. Thus

$$u_*^0 : \xi_1^+(0) \mapsto u(0),$$

defines a curve, which we call an unstable fiber denoted by $\mathcal{F}_p^+$, where $p = (J_0, \theta_0)$ is the base point, $\xi_1^+(0) \in [-r, r] \times [-r, r]$.

Let S^t denote the evolution operator of (5.27)-(5.28), then

$$S^t \mathcal{F}_p^t \subset \mathcal{F}_{S^t p}^t, \quad \forall t \leq 0.$$

That is, $\{\mathcal{F}_p^+\}_{p \in \mathcal{A}(4\delta)}$ is an invariant family of unstable fibers. The proof of the Unstable Fiber Theorem is finished. Q.E.D.

REMARK 5.11. If one replaces the base orbit $(J_*(t), \theta_*(t))$ by a general orbit for which only $\| \ \|_n$ norm is bounded, then the estimate (5.40) will not be possible. The $\| \ \|_{\sigma, n+2}$ norm of the fixed point g_* will not be bounded either. In such case, g_* may not be smooth in ϵ due to the singular perturbation.

REMARK 5.12. Smoothness of g_* in ϵ at $\epsilon = 0$ is a key point in locating homoclinic orbits as discussed in later sections. From integrable theory, information is known at $\epsilon = 0$. This key point will link "$\epsilon = 0$" information to "$\epsilon \neq 0$" studies. Only continuity in ϵ at $\epsilon = 0$ is not enough for the study. The beauty of the entire theory is reflected by the fact that although $e^{\epsilon \partial_x^2}$ is not holomorphic at $\epsilon = 0$, $e^{\epsilon \partial_x^2} g_*$ can be smooth at $\epsilon = 0$ from the right, up to certain order depending upon the regularity of g_*. This is the beauty of the singular perturbation.

5.4. Proof of the Center-Stable Manifold Theorem 5.4

Here we give the proof of the center-stable manifold theorem 5.4, proofs of other invariant manifold theorems in this chapter are easier.

5.4.1. Existence of the Center-Stable Manifold. We start with Equations (5.20)-(5.24), let

$$(5.43) \qquad v = \begin{pmatrix} J \\ \theta \\ h \\ \xi_1^- \end{pmatrix}, \quad \tilde{v} = \begin{pmatrix} J \\ h \\ \xi_1^- \end{pmatrix},$$

and let $E_n(r)$ be the tubular neighborhood of S_ω (5.18):

$$(5.44) \qquad E_n(r) = \{(J, \theta, h, \xi_1^-) \in H^n \mid \|\tilde{v}\|_n \leq r\}.$$

$E_n(r)$ is of codimension 1 in the entire phase space coordinatized by $(\xi_1^+, J, \theta, h, \xi_1^-)$.

Let $\chi \in C^\infty(R, R)$ be a "cut-off" function:

$$\chi = \begin{cases} 0, & \text{in } (-\infty, -4) \cup (4, \infty), \\ 1, & \text{in } (-2, 2). \end{cases}$$

We apply the cut-off

$$\chi_\delta = \chi(\|\tilde{v}\|_n / \delta) \chi(\xi_1^+ / \delta)$$

to Equations (5.20)-(5.24), so that the equations in a tubular neighborhood of the circle S_ω (5.18) are unchanged, and linear outside a bigger tubular neighborhood. The modified equations take the form:

$$(5.45) \qquad \dot{\xi}_1^+ = \mu_1^+ \xi_1^+ + \tilde{F}_1^+,$$

$$(5.46) \qquad v_t = Av + \tilde{F},$$

where A is given in (5.28),

$$\tilde{F}_1^+ = \chi_\delta [V_1^+ \xi_1^+ + \mathcal{N}_1^+],$$

$$\tilde{F} = (\tilde{F}_J, \tilde{F}_\theta, \tilde{F}_h, \tilde{F}_1^-)^T, \quad T = \text{transpose},$$

$$\tilde{F}_J = \chi_\delta \, \epsilon \left[-2\alpha(J + \omega^2) + 2\beta\sqrt{J + \omega^2} \cos\theta + \mathcal{R}_2^J \right],$$

$$\tilde{F}_\theta = \chi_\delta \left[-\epsilon\beta \frac{\sin\theta}{\sqrt{J + \omega^2}} + \mathcal{R}_2^\theta \right],$$

$$\tilde{F}_h = \chi_\delta [V_\epsilon h + \tilde{\mathcal{N}}],$$

$$\tilde{F}_1^- = \chi_\delta [V_1^- \xi_1^- + \mathcal{N}_1^-],$$

Equations (5.45)-(5.46) can be written in the equivalent integral equation form:

$$(5.47) \qquad \xi_1^+(t) = \xi_1^+(t_0) e^{\mu_1^+(t-t_0)} + \int_{t_0}^t e^{\mu_1^+(t-\tau)} \tilde{F}_1^+(\tau) d\tau,$$

$$(5.48) \qquad v(t) = e^{A(t-t_0)} v(t_0) + \int_{t_0}^t e^{A(t-\tau)} \tilde{F}(\tau) d\tau.$$

We introduce the following space: For $\sigma \in \left(\frac{\mu_1^+}{100}, \frac{\mu_1^+}{3} \right)$, and $n \geq 1$, let

$$\tilde{G}_{\sigma,n} = \left\{ g(t) = (\xi_1^+(t), v(t)) \,\middle|\, t \in [0,\infty), g(t) \text{ is continuous in } t \right.$$

$$\left. \text{in } H^n \text{ norm}, \|g\|_{\sigma,n} = \sup_{t \geq 0} e^{-\sigma t}[|\xi_1^+(t)| + \|v(t)\|_n] < \infty \right\}.$$

$\tilde{G}_{\sigma,n}$ is a Banach space under the norm $\| \cdot \|_{\sigma,n}$. Let $\tilde{\mathcal{A}}_{\sigma,n}(r)$ denote the closed tubular neighborhood of S_ω (5.18):

$$\tilde{\mathcal{A}}_{\sigma,n}(r) = \left\{ g(t) = (\xi_1^+(t), v(t)) \in \tilde{G}_{\sigma,n} \,\middle|\, \sup_{t \geq 0} e^{-\sigma t}[|\xi_1^+(t)| + \|\tilde{v}(t)\|_n] \leq r \right\},$$

where $\tilde{v}$ is defined in (5.43). If $g(t) \in \tilde{\mathcal{A}}_{\sigma,n}(r)$, $r < \infty$, is a solution of (5.47)-(5.48), by letting $t_0 \to +\infty$ in (5.47) and setting $t_0 = 0$ in (5.48), one has

$$(5.49) \qquad \xi_1^+(t) = \int_{+\infty}^t e^{\mu_1^+(t-\tau)} \tilde{F}_1^+(\tau) d\tau,$$

$$(5.50) \qquad v(t) = e^{At} v(0) + \int_0^t e^{A(t-\tau)} \tilde{F}(\tau) d\tau.$$

For any $g(t) \in \tilde{\mathcal{A}}_{\sigma,n}(r)$, let $\tilde{\Gamma}(g)$ be the map defined by the right hand side of (5.49)-(5.50). In contrast to the map Γ defined in (5.31)-(5.32), $\tilde{\Gamma}$ contains constant terms of order $\mathcal{O}(\epsilon)$, e.g. $\tilde{F}_J$ and $\tilde{F}_\theta$ both contain such terms. Also, $\tilde{\mathcal{A}}_{\sigma,n}(r)$ is a tubular neighborhood of the circle S_ω (5.18) instead of the ball $\mathcal{B}_{\sigma,n}(r)$ for Γ. Fortunately, these facts will not create any difficulty in showing $\tilde{\Gamma}$ is a contraction on $\tilde{\mathcal{A}}_{\sigma,n}(r)$.

For any $n \geq 1$ and $\epsilon < \delta^2$, and δ and r are small enough, $\tilde{F}_1^+$ and $\tilde{F}$ are Lipschitz in g with small Lipschitz constants. $\tilde{\Gamma}$ has a unique fixed point $\tilde{g}_*$ in $\tilde{\mathcal{A}}_{\sigma,n}(r)$, following from standard arguments.

5.4.2. Regularity of the Center-Stable Manifold in ϵ. For the regularity of $\tilde{g}_*$ with respect to $(\epsilon, \alpha, \beta, \omega, v(0))$, the most difficult one is of course with respect to ϵ due to the singular perturbation. Formally differentiating $\tilde{g}_*$ in (5.49)-(5.50) with respect to ϵ, one gets

$$(5.51) \qquad \xi_{1,\epsilon}^+(t) = \int_{+\infty}^{t} e^{\mu_1^+(t-\tau)} \left[\partial_{\xi_1^+} \tilde{F}_1^+ \cdot \xi_{1,\epsilon}^+ + \partial_v \tilde{F}_1^+ \cdot v_\epsilon \right] (\tau) d\tau + \tilde{R}_1^+(t),$$

$$(5.52) \qquad v(t) = \int_0^t e^{A(t-\tau)} \left[\partial_{\xi_1^+} \tilde{F} \cdot \xi_{1,\epsilon}^+ + \partial_v \tilde{F} \cdot v_\epsilon \right] (\tau) d\tau + \tilde{R}(t),$$

where

$$(5.53) \quad \tilde{R}_1^+(t) = \int_{+\infty}^{t} \mu_{1,\epsilon}^+(t-\tau) e^{\mu_1^+(t-\tau)} \tilde{F}_1^+(\tau) d\tau + \int_{+\infty}^{t} e^{\mu_1^+(t-\tau)} \partial_\epsilon \tilde{F}_1^+(\tau) d\tau,$$

$$(5.54) \quad \tilde{R}(t) = t A_\epsilon e^{At} v(0) + \int_0^t (t-\tau) A_\epsilon e^{A(t-\tau)} \tilde{F}(\tau) d\tau + \int_0^t e^{A(t-\tau)} \partial_\epsilon \tilde{F}(\tau) d\tau,$$

and $\mu_{1,\epsilon}^+$ and A_ϵ are given in (5.37)-(5.39). The troublesome terms are the ones containing A_ϵ in (5.54). These terms can be handled in the same way as in the Proof of the Unstable Fiber Theorem. The crucial fact utilized is that if $v(0) \in H^{n_1}$, then $\tilde{g}_*$ is the unique fixed point of $\tilde{\Gamma}$ in both $\tilde{G}_{\sigma,n_1}$ and $\tilde{G}_{\sigma,n_2}$ for any $n_2 \leq n_1$.

REMARK 5.13. In the Proof of the Unstable Fiber Theorem, the arbitrary initial datum in (5.31)-(5.32) is $\xi_1^+(0)$ which is a scalar. Here the arbitrary initial datum in (5.49)-(5.50) is $v(0)$ which is a function of x. If $v(0) \in H^{n_2}$ but not H^{n_1} for some $n_1 > n_2$, then $\tilde{g}_* \notin \tilde{G}_{\sigma,n_1}$, in contrast to the case of (5.31)-(5.32) where $g_* \in G_{\sigma,n}$ for any fixed $n \geq 1$. The center-stable manifold W_n^{cs} stated in the Center-Stable Manifold Theorem will be defined through $v(0)$. This already illustrates why W_n^{cs} has the regularity in ϵ as stated in the theorem.

We have

$$\|\tilde{R}_1^+\|_{\sigma,n} \leq \tilde{C}_1,$$

$$\|\tilde{R}\|_{\sigma,n} \leq \tilde{C}_2 \|\tilde{g}_*\|_{\sigma,n+2} + \tilde{C}_3,$$

for $\tilde{g}_* \in \tilde{\mathcal{A}}_{\sigma,n+2}(r)$, where $\tilde{C}_j$ $(j = 1, 2, 3)$ are constants depending in particular upon the cut-off in $\tilde{F}_1^+$ and $\tilde{F}$. Let $\tilde{\Gamma}'$ denote the linear map defined by the right hand sides of (5.51)-(5.52). If $v(0) \in H^{n+2}$ and $\tilde{g}_* \in \tilde{\mathcal{A}}_{\sigma,n+2}(r)$, standard argument shows that $\tilde{\Gamma}'$ is a contraction map on a closed ball in $L(R, \tilde{G}_{\sigma,n})$. Thus $\tilde{\Gamma}'$ has a unique fixed point $\tilde{g}_{*,\epsilon}$. Furthermore, if $v(0) \in H^{n+4}$ and $\tilde{g}_* \in \tilde{\mathcal{A}}_{\sigma,n+4}(r)$, one has that $\tilde{g}_{*,\epsilon}$ is indeed the derivative of $\tilde{g}_*$ in ϵ, following the same argument as in the Proof of the Unstable Fiber Theorem. Here one may be able to replace the requirement $v(0) \in H^{n+4}$ and $\tilde{g}_* \in \tilde{\mathcal{A}}_{\sigma,n+4}(r)$ by just $v(0) \in H^{n+2}$ and $\tilde{g}_* \in \tilde{\mathcal{A}}_{\sigma,n+2}(r)$. But we are not interested in sharper results, and the current result is sufficient for our purpose.

DEFINITION 5.14. For any $v(0) \in E_n(r)$ where r is sufficiently small and $E_n(r)$ is defined in (5.44), let $\tilde{g}_*(t) = (\xi_1^+(t), v(t))$ be the fixed point of $\tilde{\Gamma}$ in $\tilde{G}_{\sigma,n}$, where one has

$$\xi_1^+(0) = \int_{+\infty}^0 e^{\mu_1^+(t-\tau)} \tilde{F}_1^+(\tau)d\tau,$$

which depend upon $v(0)$. Thus

$$\xi_*^+ : v(0) \mapsto \xi_1^+(0),$$

defines a codimension 1 surface, which we call center-stable manifold denoted by W_n^{cs}.

The regularity of the fixed point $\tilde{g}_*$ immediately implies the regularity of W_n^{cs}. We have sketched the proof of the most difficult regularity, i.e. with respect to ϵ. Uniform boundedness of $\partial_\epsilon \xi_*^+$ in $v(0) \in E_{n+4}(r)$ and $\epsilon \in [0, \epsilon_0)$, is obvious. Other parts of the detailed proof is completely standard. We have that W_n^{cs} is a C^1 locally invariant submanifold which is C^1 in (α, β, ω). W_n^{cs} is C^1 in ϵ at point in the subset W_{n+4}^{cs}. Q.E.D.

REMARK 5.15. Let S^t denote the evolution operator of the perturbed nonlinear Schrödinger equation (5.2). The proofs of the Unstable Fiber Theorem and the Center-Stable Manifold Theorem also imply the following: S^t is a C^1 map on H^n for any fixed $t > 0$, $n \geq 1$. S^t is also C^1 in (α, β, ω). S^t is C^1 in ϵ as a map from H^{n+4} to H^n for any fixed $n \geq 1$, $\epsilon \in [0, \epsilon_0)$, $\epsilon_0 > 0$.

5.5. Perturbed Davey-Stewartson II Equations

Invariant manifold results in Sections 5.2 and 5.1 also hold for perturbed Davey-Stewartson II equations [132],

$$iq_t = \Upsilon q + \left[2(|q|^2 - \omega^2) + u_y\right]q + i\epsilon f \ ,$$

$$\Delta u = -4\partial_y |q|^2 \ ,$$

where q is a complex-valued function of the three variables (t, x, y), u is a real-valued function of the three variables (t, x, y), $\Upsilon = \partial_{xx} - \partial_{yy}$, $\Delta = \partial_{xx} + \partial_{yy}$, $\omega > 0$ is a constant, and f is the perturbation. We also consider periodic boundary conditions.

Under singular perturbation

$$f = \Delta q - \alpha q + \beta \ ,$$

where $\alpha > 0$, $\beta > 0$ are constants, Theorems 5.3 and 5.4 hold for the perturbed Davey-Stewartson II equations [132].

When the singular perturbation Δ is mollified into a bounded Fourier multiplier

$$\hat{\Delta} q = - \sum_{k \in Z^2} \beta_k |k|^2 \tilde{q}_k \cos k_1 x \cos k_2 y \ ,$$

in the case of periods $(2\pi, 2\pi)$,

$$\beta_k = 1, \quad |k| \leq N, \quad \beta_k = |k|^{-2}, \quad |k| > N,$$

for some large N, $|k|^2 = k_1^2 + k_2^2$, Theorems 5.1 and 5.2 hold for the perturbed Davey-Stewartson II equations [132].

5.6. General Overview

For discrete systems, i.e., the flow is given by a map, it is more convenient to use Hadamard's method to prove invariant manifold and fiber theorems [84]. Even for continuous systems, Hadamard's method was often utilized [64]. On the other hand, Perron's method provides shorter proofs. It involves manipulation of integral equations. This method should be a favorite of analysts. Hadamard's method deals with graph transform. The proof is often much longer, with clear geometric intuitions. It should be a favorite of geometers. For finite-dimensional continuous systems, N. Fenichel proved persistence of invariant manifolds under C^1 perturbations of flow in a very general setting [64]. He then went on to prove the fiber theorems in [65] [66] also in this general setting. Finally, he applied this general machinery to a general system of ordinary differential equations [67]. As a result, Theorems 5.1 and 5.2 hold for the following perturbed discrete cubic nonlinear Schrödinger equations [138],

$$i\dot{q}_n = \frac{1}{h^2}\left[q_{n+1} - 2q_n + q_{n-1}\right] + |q_n|^2(q_{n+1} + q_{n-1}) - 2\omega^2 q_n$$

$$(5.55) \qquad\qquad +i\epsilon\left[-\alpha q_n + \frac{1}{h^2}(q_{n+1} - 2q_n + q_{n-1}) + \beta\right],$$

where $i = \sqrt{-1}$, q_n's are complex variables,

$$q_{n+N} = q_n, \quad \text{(periodic condition)}; \quad \text{and} \quad q_{-n} = q_n, \quad \text{(even condition)};$$

$h = \frac{1}{N}$, and

$$N\tan\frac{\pi}{N} < \omega < N\tan\frac{2\pi}{N}, \quad \text{for } N > 3,$$

$$3\tan\frac{\pi}{3} < \omega < \infty, \quad \text{for } N = 3.$$

$$\epsilon \in [0, \epsilon_1), \ \alpha \ (> 0), \ \beta \ (> 0) \text{ are constants.}$$

This is a $2(M + 1)$ dimensional system, where

$$M = N/2, \quad (N \text{ even}); \quad \text{and} \quad M = (N - 1)/2, \quad (N \text{ odd}).$$

This system is a finite-difference discretization of the perturbed NLS (5.2).

For a general system of ordinary differential equations, Kelley [103] used the Perron's method to give a very short proof of the classical unstable, stable, and center manifold theorem. This paper is a good starting point of reading upon Perron's method. In the book [84], Hadamard's method is mainly employed. This book is an excellent starting point for a comprehensive reading on invariant manifolds.

There have been more and more invariant manifold results for infinite dimensional systems [140]. For the employment of Perron's method, we refer the readers to [42]. For the employment of Hadamard's method, we refer the readers to [18] [19] [20] which are terribly long papers.

CHAPTER 6

Homoclinic Orbits

In terms of proving the existence of a homoclinic orbit, the most common tool is the so-called Melnikov integral method [158] [12]. This method was subsequently developed by Holmes and Marsden [76], and most recently by Wiggins [201]. For partial differential equations, this method was mainly developed by Li et al. [136] [124] [138] [128].

There are two derivations for the Melnikov integrals. One is the so-called geometric argument [76] [201] [138] [136] [124]. The other is the so-called Liapunov-Schmitt argument [41] [40]. The Liapunov-Schmitt argument is a fixed-point type argument which directly leads to the existence of a homoclinic orbit. The condition for the existence of a fixed point is the Melnikov integral. The geometric argument is a signed distance argument which applies to more general situations than the Liapunov-Schmitt argument. It turns out that the geometric argument is a much more powerful machinary than the Liapunov-Schmitt argument. In particular, the geometric argument can handle geometric singular perturbation problems. I shall also mention an interesting derivation in [12].

In establishing the existence of homoclinic orbits in high dimensions, one often needs other tools besides the Melnikov analysis. For example, when studying orbits homoclinic to fixed points created through resonances in $(n \geq 4)$-dimensional near-integrable systems, one often needs tools like Fenichel fibers, as presented in previous chapter, to set up geometric measurements for locating such homoclinci orbits. Such homoclinic orbits often have a geometric singular perturbation nature. In such cases, the Liapunov-Schmitt argument can not be applied. For such works on finite dimensional systems, see for example [108] [109] [138]. For such works on infinite dimensional systems, see for example [136] [124].

6.1. Silnikov Homoclinic Orbits in NLS Under Regular Perturbations

We continue from section 5.1 and consider the regularly perturbed nonlinear Schrödinger (NLS) equation (5.1). The following theorem was proved in [136].

THEOREM 6.1. *There exists a $\epsilon_0 > 0$, such that for any $\epsilon \in (0, \epsilon_0)$, there exists a codimension 1 surface in the external parameter space $(\alpha, \beta, \omega) \in \mathbb{R}^+ \times \mathbb{R}^+ \times \mathbb{R}^+$ where $\omega \in (\frac{1}{2}, 1)$, and $\alpha\omega < \beta$. For any (α, β, ω) on the codimension 1 surface, the regularly perturbed nonlinear Schrödinger equation (5.1) possesses a symmetric pair of Silnikov homoclinic orbits asymptotic to a saddle Q_ϵ. The codimension 1 surface has the approximate representation given by $\alpha = 1/\kappa(\omega)$, where $\kappa(\omega)$ is plotted in Figure 1.*

The proof of this theorem is easier than that given in later sections. To prove the theorem, one starts from the invariant plane

$$\Pi = \{q \mid \partial_x q = 0\}.$$

55

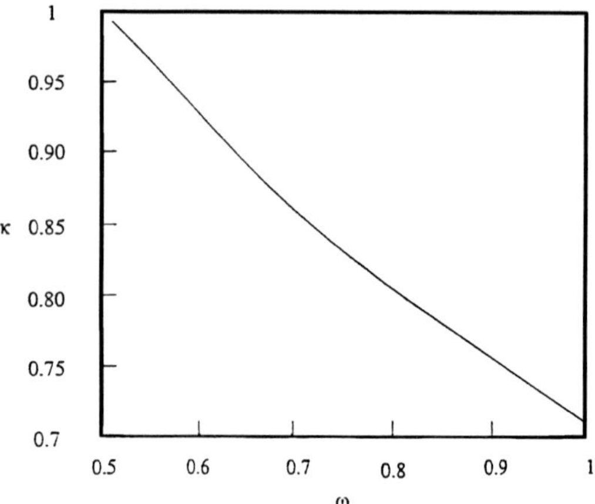

FIGURE 1. The graph of $\kappa(\omega)$.

On Π, there is a saddle $Q_\epsilon = \sqrt{I}e^{i\theta}$ to which the symmetric pair of Silnikov homoclinic orbits will be asymptotic to, where

$$(6.1) \qquad I = \omega^2 - \epsilon \frac{1}{2\omega}\sqrt{\beta^2 - \alpha^2\omega^2} + \cdots, \qquad \cos\theta = \frac{\alpha\sqrt{I}}{\beta}, \qquad \theta \in (0, \frac{\pi}{2}).$$

Its eigenvalues are

$$(6.2) \qquad \lambda_n^\pm = -\epsilon[\alpha + \xi_n n^2] \pm 2\sqrt{(\frac{n^2}{2} + \omega^2 - I)(3I - \omega^2 - \frac{n^2}{2})},$$

where $n = 0, 1, 2, \cdots$, $\omega \in (\frac{1}{2}, 1)$, $\xi_n = 1$ when $n \leq N$, $\xi_n = 8n^{-2}$ when $n > N$, for some fixed large N, and I is given in (6.1). The crucial points to notice are: (1). only λ_0^+ and λ_1^+ have positive real parts, $\text{Re}\{\lambda_0^+\} < \text{Re}\{\lambda_1^+\}$; (2). all the other eigenvalues have negative real parts among which the absolute value of $\text{Re}\{\lambda_2^+\} = \text{Re}\{\lambda_2^-\}$ is the smallest; (3). $|\text{Re}\{\lambda_2^+\}| < \text{Re}\{\lambda_0^+\}$. Actually, items (2) and (3) are the main characteristics of Silnikov homoclinic orbits.

The unstable manifold $W^u(Q_\epsilon)$ of Q_ϵ has a fiber representation given by Theorem 5.2. The Melnikov measurement measures the signed distance between $W^u(Q_\epsilon)$ and the center-stable manifold W_ϵ^{cs} proved in Thoerem 5.1. By virtue of the Fiber Theorem 5.2, one can show that, to the leading order in ϵ, the signed distance is given by the Melnikov integral

$$M = \int_{-\infty}^{+\infty} \int_0^{2\pi} [\partial_q F_1(q_0(t))(\hat{\partial}_x^2 q_0(t) - \alpha q_0(t) + \beta)$$
$$+ \partial_{\bar{q}} F_1(q_0(t))(\hat{\partial}_x^2 \overline{q_0(t)} - \alpha\overline{q_0(t)} + \beta)]dxdt,$$

where $q_0(t)$ is given in section 3.1, equation (3.14); and $\partial_q F_1$ and $\partial_{\bar{q}} F_1$ are given in section 4.1, equation (4.15). The zero of the signed distance implies the existence of an orbit in $W^u(Q_\epsilon) \cap W_\epsilon^{cs}$. The stable manifold $W^s(Q_\epsilon)$ of Q_ϵ is a codimension 1 submanifold in W_ϵ^{cs}. To locate a homoclinic orbit, one needs to set up a second measurement measuring the signed distance between the orbit in $W^u(Q_\epsilon) \cap W_\epsilon^{cs}$ and $W^s(Q_\epsilon)$ inside W_ϵ^{cs}. To set up this signed distance, first one can rather easily track the (perturbed) orbit by an unperturbed orbit to an $\mathcal{O}(\epsilon)$ neighborhood of

Π, then one needs to prove the size of $W^s(Q_\epsilon)$ to be $\mathcal{O}(\epsilon^\nu)$ ($\nu < 1$) with normal form transform. To the leading order in ϵ, the zero of the second signed distance is given by

$$\beta \cos \gamma = \frac{\alpha \omega (\Delta \gamma)}{2 \sin \frac{\Delta \gamma}{2}} ,$$

where $\Delta \gamma = -4\vartheta_0$ and ϑ_0 is given in (3.14). To the leading order in ϵ, the common zero of the two second signed distances satisfies $\alpha = 1/\kappa(\omega)$, where $\kappa(\omega)$ is plotted in Figure 1. Then the claim of the theorem is proved by virtue of the implicit function theorem. For rigorous details, see [136]. In the singular perturbation case as discussed in next section, the rigorous details are given in later sections.

6.2. Silnikov Homoclinic Orbits in NLS Under Singular Perturbations

We continue from section 5.2 and consider the singularly perturbed nonlinear Schrödinger (NLS) equation (5.2). The following theorem was proved in [124].

THEOREM 6.2. *There exists a $\epsilon_0 > 0$, such that for any $\epsilon \in (0, \epsilon_0)$, there exists a codimension 1 surface in the external parameter space $(\alpha, \beta, \omega) \in \mathbb{R}^+ \times \mathbb{R}^+ \times \mathbb{R}^+$ where $\omega \in (\frac{1}{2}, 1)/S$, S is a finite subset, and $\alpha\omega < \beta$. For any (α, β, ω) on the codimension 1 surface, the singularly perturbed nonlinear Schrödinger equation (5.2) possesses a symmetric pair of Silnikov homoclinic orbits asymptotic to a saddle Q_ϵ. The codimension 1 surface has the approximate representation given by $\alpha = 1/\kappa(\omega)$, where $\kappa(\omega)$ is plotted in Figure 1.*

In this singular perturbation case, the persistence and fiber theorems are given in section 5.2, Theorems 5.4 and 5.3. The normal form transform for proving the size estimate of the stable manifold $W^s(Q_\epsilon)$ is still achievable. The proof of the theorem is also completed through two measurements: the Melnikov measurement and the second measurement.

6.3. The Melnikov Measurement

6.3.1. Dynamics on an Invariant Plane. The 2D subspace Π,

(6.3) $$\Pi = \{q \mid \partial_x q = 0\},$$

is an invariant plane under the flow (5.2). The governing equation in Π is

(6.4) $$i\dot{q} = 2[|q|^2 - \omega^2]q + i\epsilon[-\alpha q + \beta],$$

where $\cdot = \frac{d}{dt}$. Dynamics of this equation is shown in Figure 2. Interesting dynamics is created through resonance in the neighborhood of the circle S_ω:

(6.5) $$S_\omega = \{q \in \Pi \mid |q| = \omega\}.$$

When $\epsilon = 0$, S_ω consists of fixed points. To explore the dynamics in this neighborhood better, one can make a series of changes of coordinates. Let $q = \sqrt{I}e^{i\theta}$, then (6.4) can be rewritten as

(6.6) $$\dot{I} = \epsilon(-2\alpha I + 2\beta\sqrt{I}\cos\theta) ,$$

(6.7) $$\dot{\theta} = -2(I - \omega^2) - \epsilon\beta\frac{\sin\theta}{\sqrt{I}} .$$

There are three fixed points:

(1) The focus O_ϵ in the neighborhood of the origin,

$$(6.8) \qquad \begin{cases} I = \epsilon^2 \frac{\beta^2}{4\omega^4} + \cdots, \\ \cos\theta = \frac{\alpha\sqrt{I}}{\beta}, \end{cases} \qquad \theta \in \left(0, \tfrac{\pi}{2}\right).$$

Its eigenvalues are

$$(6.9) \qquad \mu_{1,2} = \pm i\sqrt{4(\omega^2 - I)^2 - 4\epsilon\sqrt{I}\beta\sin\theta} - \epsilon\alpha,$$

where I and θ are given in (6.8).

(2) The focus P_ϵ in the neighborhood of S_ω (6.5),

$$(6.10) \qquad \begin{cases} I = \omega^2 + \epsilon\frac{1}{2\omega}\sqrt{\beta^2 - \alpha^2\omega^2} + \cdots, \\ \cos\theta = \frac{\alpha\sqrt{I}}{\beta}, \end{cases} \qquad \theta \in \left(-\tfrac{\pi}{2}, 0\right).$$

Its eigenvalues are

$$(6.11) \qquad \mu_{1,2} = \pm i\sqrt{\epsilon}\sqrt{-4\sqrt{I}\beta\sin\theta + \epsilon\left(\frac{\beta\sin\theta}{\sqrt{I}}\right)^2} - \epsilon\alpha,$$

where I and θ are given in (6.10).

(3) The saddle Q_ϵ in the neighborhood of S_ω (6.5),

$$(6.12) \qquad \begin{cases} I = \omega^2 - \epsilon\frac{1}{2\omega}\sqrt{\beta^2 - \alpha^2\omega^2} + \cdots, \\ \cos\theta = \frac{\alpha\sqrt{I}}{\beta}, \end{cases} \qquad \theta \in \left(0, \tfrac{\pi}{2}\right).$$

Its eigenvalues are

$$(6.13) \qquad \mu_{1,2} = \pm\sqrt{\epsilon}\sqrt{4\sqrt{I}\beta\sin\theta - \epsilon\left(\frac{\beta\sin\theta}{\sqrt{I}}\right)^2} - \epsilon\alpha,$$

where I and θ are given in (6.12).

Now focus our attention to order $\sqrt{\epsilon}$ neighborhood of S_ω (6.5) and let

$$J = I - \omega^2, \quad J = \sqrt{\epsilon}j, \quad \tau = \sqrt{\epsilon}t,$$

we have

$$(6.14) \qquad j' = 2\left[-\alpha(\omega^2 + \sqrt{\epsilon}j) + \beta\sqrt{\omega^2 + \sqrt{\epsilon}j}\cos\theta\right],$$

$$(6.15) \qquad \theta' = -2j - \sqrt{\epsilon}\beta\frac{\sin\theta}{\sqrt{\omega^2 + \sqrt{\epsilon}j}},$$

where $' = \frac{d}{d\tau}$. To leading order, we get

$$(6.16) \qquad j' = 2[-\alpha\omega^2 + \beta\omega\cos\theta],$$

$$(6.17) \qquad \theta' = -2j.$$

There are two fixed points which are the counterparts of P_ϵ and Q_ϵ (6.10) and (6.12):

(1) The center P_*,

$$(6.18) \qquad j = 0, \quad \cos\theta = \frac{\alpha\omega}{\beta}, \quad \theta \in \left(-\frac{\pi}{2}, 0\right).$$

Its eigenvalues are

$$(6.19) \qquad \mu_{1,2} = \pm i2\sqrt{\omega}(\beta^2 - \alpha^2\omega^2)^{\frac{1}{4}}.$$

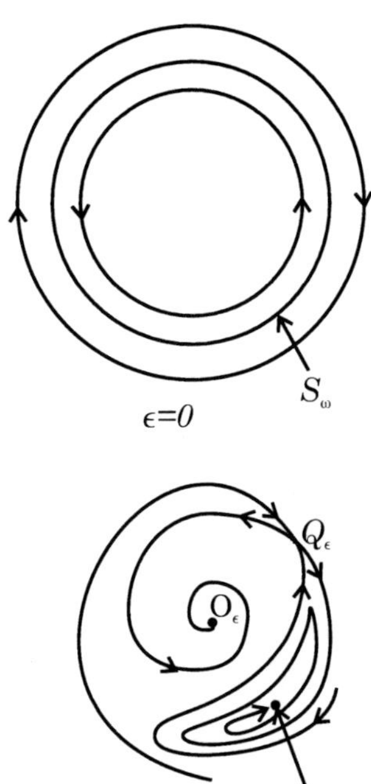

FIGURE 2. Dynamics on the invariant plane Π.

(2) The saddle Q_*,

$$
(6.20) \qquad j = 0, \quad \cos\theta = \frac{\alpha\omega}{\beta}, \quad \theta \in \left(0, \frac{\pi}{2}\right).
$$

Its eigenvalues are

$$
(6.21) \qquad \mu_{1,2} = \pm 2\sqrt{\omega}(\beta^2 - \alpha^2\omega^2)^{\frac{1}{4}}.
$$

In fact, (6.16) and (6.17) form a Hamiltonian system with the Hamiltonian

$$
(6.22) \qquad \mathcal{H} = j^2 + 2\omega(-\alpha\omega\theta + \beta\sin\theta).
$$

Connecting to Q_* is a fish-like singular level set of $\mathcal{H}$, which intersects the axis $j = 0$ at Q_* and $\hat{Q} = (0, \hat{\theta})$,

$$
(6.23) \qquad \alpha\omega(\hat{\theta} - \theta_*) = \beta(\sin\hat{\theta} - \sin\theta_*), \quad \hat{\theta} \in \left(-\frac{3\pi}{2}, 0\right),
$$

where θ_* is given in (6.20). See Figure 3 for an illustration of the dynamics of (6.14)-(6.17). For later use, we define a piece of each of the stable and unstable manifolds of Q_*,

$$
j = \phi_*^u(\theta), \quad j = \phi_*^s(\theta), \quad \theta \in [\hat{\theta} + \hat{\delta}, \theta_* + 2\pi],
$$

for some small $\hat{\delta} > 0$, and

$$
(6.24) \qquad
\begin{aligned}
\phi_*^u(\theta) &= -\frac{\theta - \theta_*}{|\theta - \theta_*|} \sqrt{2\omega[\alpha\omega(\theta - \theta_*) - \beta(\sin\theta - \sin\theta_*)]}, \\
\phi_*^s(\theta) &= -\phi_*^u(\theta).
\end{aligned}
$$

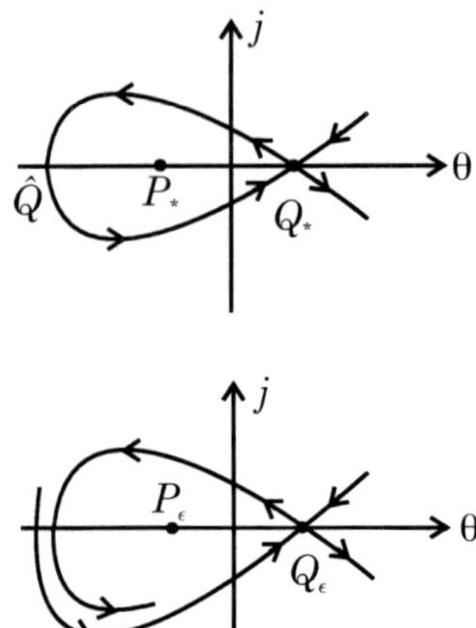

FIGURE 3. The fish-like dynamics in the neighborhood of the resonant circle S_ω.

$\phi^u_*(\theta)$ and $\phi^s_*(\theta)$ perturb smoothly in θ and $\sqrt{\epsilon}$ into $\phi^u_{\sqrt{\epsilon}}$ and $\phi^s_{\sqrt{\epsilon}}$ for (6.14) and (6.15).

The homoclinic orbit to be located will take off from Q_ϵ along its unstable curve, flies away from and returns to Π, lands near the stable curve of Q_ϵ and approaches Q_ϵ spirally.

6.3.2. A Signed Distance. Let p be any point on $\phi^u_{\sqrt{\epsilon}}$ (6.24) which is the unstable curve of Q_ϵ in Π (6.3). Let $q_\epsilon(0)$ and $q_0(0)$ be any two points on the unstable fibers $\mathcal{F}^+_p \mid_\epsilon$ and $\mathcal{F}^+_p \mid_{\epsilon=0}$, with the same ξ^+_1 coordinate. See Figure 4 for an illustration. By the Unstable Fiber Theorem 5.3, $\mathcal{F}^+_p$ is C^1 in ϵ for $\epsilon \in [0, \epsilon_0)$, $\epsilon_0 > 0$, thus

$$\|q_\epsilon(0) - q_0(0)\|_{n+8} \le C\epsilon.$$

The key point here is that $\mathcal{F}^+_p \subset H^s$ for any fixed $s \ge 1$. By Remark 5.15, the evolution operator of the perturbed NLS equation (5.2) S^t is C^1 in ϵ as a map from H^{n+4} to H^n for any fixed $n \ge 1$, $\epsilon \in [0, \epsilon_0)$, $\epsilon_0 > 0$. Also S^t is a C^1 map on H^n for any fixed $t > 0$, $n \ge 1$. Thus

$$\|q_\epsilon(T) - q_0(T)\|_{n+4} = \|S^T(q_\epsilon(0)) - S^T(q_0(0))\|_{n+4} \le C_1\epsilon,$$

where $T > 0$ is large enough so that

$$q_0(T) \in W^{cs}_{n+4} \mid_{\epsilon=0} .$$

Our goal is to determine when $q_\epsilon(T) \in W^{cs}_n$ through Melnikov measurement. Let $q_\epsilon(T)$ and $q_0(T)$ have the coordinate expressions

(6.25) $$q_\epsilon(T) = (\xi^{+,\epsilon}_1, v_\epsilon), \quad q_0(T) = (\xi^{+,0}_1, v_0).$$

Let $\tilde{q}_\epsilon(T)$ be the unique point on W^{cs}_{n+4}, which has the same v-coordinate as $q_\epsilon(T)$,

$$\tilde{q}_\epsilon(T) = (\tilde{\xi}^{+,\epsilon}_1, v_\epsilon) \in W^{cs}_{n+4}.$$

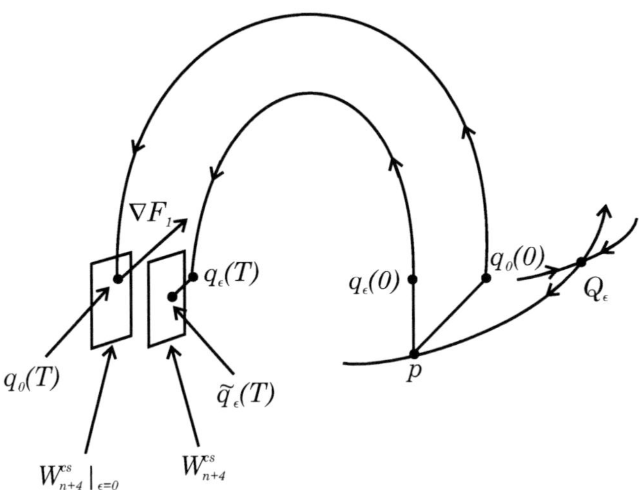

FIGURE 4. The Melnikov measurement.

By the Center-Stable Manifold Theorem, at points in the subset W^{cs}_{n+4}, W^{cs}_n is C^1 smooth in ϵ for $\epsilon \in [0, \epsilon_0)$, $\epsilon_0 > 0$, thus

$$(6.26) \qquad \|q_\epsilon(T) - \tilde{q}_\epsilon(T)\|_n \le C_2 \epsilon.$$

Also our goal now is to determine when the signed distance

$$\xi_1^{+,\epsilon} - \tilde{\xi}_1^{+,\epsilon},$$

is zero through Melnikov measurement. Equivalently, one can define the signed distance

$$d_1 = \langle \nabla F_1(q_0(T)), q_\epsilon(T) - \tilde{q}_\epsilon(T) \rangle$$
$$= \partial_q F_1(q_0(T))(q_\epsilon(T) - \tilde{q}_\epsilon(T)) + \partial_{\bar{q}} F_1(q_0(T))(q_\epsilon(T) - \tilde{q}_\epsilon(T))^-,$$

where ∇F_1 is given in (4.17), and $q_0(t)$ is the homoclinic orbit given in (3.14). In fact, $q_\epsilon(t)$, $\tilde{q}_\epsilon(t)$, $q_0(t) \in H^n$, for any fixed $n \ge 1$. The rest of the derivation for Melnikov integrals is completely standard. For details, see [136] [138].

$$(6.27) \qquad d_1 = \epsilon M_1 + o(\epsilon),$$

where

$$M_1 = \int_{-\infty}^{+\infty} \int_0^{2\pi} [\partial_q F_1(q_0(t))(\partial_x^2 q_0(t) - \alpha q_0(t) + \beta)$$
$$+ \partial_{\bar{q}} F_1(q_0(t))(\partial_x^2 \overline{q_0(t)} - \alpha \overline{q_0(t)} + \beta)]dxdt,$$

where again $q_0(t)$ is given in (3.14), and $\partial_q F_1$, and $\partial_{\bar{q}} F_1$ are given in (4.17).

6.4. The Second Measurement

6.4.1. The Size of the Stable Manifold of Q_ϵ. Assume that the Melnikov measurement is successful, that is, the orbit $q_\epsilon(t)$ is in the intersection of the unstable manifold of Q_ϵ and the center-stable manifold $W^u(Q_\epsilon) \cap W^{cs}_n$. The goal of the second measurement is to determine when $q_\epsilon(t)$ is also in the codimension 2 stable manifold of Q_ϵ, $W_n^s(Q_\epsilon)$, in H^n. The existence of $W_n^s(Q_\epsilon)$ follows from standard stable manifold theorem. $W_n^s(Q_\epsilon)$ can be visualized as a codimension 1 wall in W_n^{cs} with base curve $\phi_{\sqrt{\epsilon}}^u$ (6.24) in Π.

THEOREM 6.3. [124] *The size of $W_n^s(Q_\epsilon)$ off Π is of order $\mathcal{O}(\sqrt{\epsilon})$ for $\omega \in \left(\frac{1}{2}, 1\right)/S$, where S is a finite subset.*

Starting from the system (5.7)-(5.9), one can only get the size of $W_n^s(Q_\epsilon)$ off Π to be $\mathcal{O}(\epsilon)$ from the standard stable manifold theorem. This is not enough for the second measurement. An estimate of order $\mathcal{O}(\epsilon^\kappa)$, $\kappa < 1$ can be achieved if the quadratic term $\mathcal{N}_2$ (5.14) in (5.9) can be removed through a normal form transformation. Such a normal form transformation has been achieved [124], see also the later section on Normal Form Transforms.

6.4.2. An Estimate. From the explicit expression of $q_0(t)$ (3.14), we know that $q_0(t)$ approaches Π at the rate $\mathcal{O}(e^{-\sqrt{4\omega^2-1}t})$. Thus

$$(6.28) \qquad \text{distance}\left\{q_0(T + \frac{1}{\mu}|\ln \epsilon|), \Pi\right\} < C\epsilon.$$

LEMMA 6.4. *For all $t \in \left[T, T + \frac{1}{\mu}|\ln \epsilon|\right]$,*

$$(6.29) \qquad \|q_\epsilon(t) - q_0(t)\|_n \leq \tilde{C}_1 \epsilon |\ln \epsilon|^2,$$

where $\tilde{C}_1 = \tilde{C}_1(T)$.

Proof: We start with the system (5.49)-(5.50). Let

$$q_\epsilon(t) = (\xi_1^{+,\epsilon}(t), J^\epsilon(t), \theta^\epsilon(t), h^\epsilon(t), \xi_1^{-,\epsilon}(t)),$$
$$q_0(t) = (\xi_1^{+,0}(t), J^0(t), \theta^0(t), h^0(t), \xi_1^{-,0}(t)).$$

Let $T_1(> T)$ be a time such that

$$(6.30) \qquad \|q_\epsilon(t) - q_0(t)\|_n \leq \tilde{C}_2 \epsilon |\ln \epsilon|^2,$$

for all $t \in [T, T_1]$, where $\tilde{C}_2 = \tilde{C}_2(T)$ is independent of ϵ. From (6.26), such a T_1 exists. The proof will be completed through a continuation argument. For $t \in [T, T_1]$,

$$(|\xi_1^{+,0}(t)| + |\xi_1^{-,0}(t)|) + \|h^0(t)\|_n \leq C_3 r e^{-\frac{1}{2}\mu(t-T)}, \qquad |J^0(t)| \leq C_4\sqrt{\epsilon},$$

$$(6.31) \quad |J^\epsilon(t)| \leq |J^0(t)| + |J^\epsilon(t) - J^0(t)| \leq |J^0(t)| + \tilde{C}_2 \epsilon |\ln \epsilon|^2 \leq C_5\sqrt{\epsilon},$$

$$(|\xi_1^{+,\epsilon}(t)| + |\xi_1^{-,\epsilon}(t)|) + \|h^\epsilon(t)\|_n \leq C_3 r e^{-\frac{1}{2}\mu(t-T)} + \tilde{C}_2 \epsilon |\ln \epsilon|^2,$$

where r is small. Since actually $q_\epsilon(t)$, $q_0(t) \in H^n$ for any fixed $n \geq 1$, by Theorem 5.4,

$$(6.32) \qquad |\xi_1^{+,\epsilon}(t) - \xi_1^{+,0}(t)| \leq C_6 \|v_\epsilon(t) - v_0(t)\|_n + C_7\epsilon,$$

whenever $v_\epsilon(t)$, $v_0(t) \in E_{n+4}(r)$, where $v_\epsilon(T) = v_\epsilon$ and $v_0(T) = v_0$ are defined in (6.25). Thus we only need to estimate $\|v_\epsilon(t) - v_0(t)\|_n$. From (5.50), we have for $t \in [T, T_1]$ that

$$(6.33) \qquad v(t) = e^{A(t-T)}v(T) + \int_T^t e^{A(t-\tau)}\tilde{F}(\tau)d\tau.$$

Let $\Delta v(t) = v_\epsilon(t) - v_0(t)$. Then

$$\Delta v(t) = [e^{A(t-T)} - e^{A|_{\epsilon=0}(t-T)}]v_0(T) + e^{A(t-T)}\Delta v(T)$$

$$(6.34) \qquad + \int_T^t e^{A(t-\tau)}[\tilde{F}(\tau) - \tilde{F}(\tau)|_{\epsilon=0}]d\tau$$

$$+ \int_T^t [e^{A(t-\tau)} - e^{A|_{\epsilon=0}(t-\tau)}] \tilde{F}(\tau)|_{\epsilon=0}d\tau.$$

By the condition (6.31), we have for $t \in [T, T_1]$ that

$$(6.35) \qquad \|\tilde{F}(t) - \tilde{F}(t)|_{\epsilon=0}\|_n \le [C_8\sqrt{\epsilon} + C_9 r e^{-\frac{1}{2}\mu(t-T)}]\epsilon|\ln\epsilon|^2.$$

Then

$$(6.36) \qquad \|\Delta v(t)\|_n \le C_{10}\epsilon(t-T) + C_{11}r\epsilon|\ln\epsilon|^2 + C_{12}\sqrt{\epsilon}(t-T)^2\epsilon|\ln\epsilon|^2.$$

Thus by the continuation argument, for $t \in [T, T + \frac{1}{\mu}|\ln\epsilon|]$, there is a constant $\hat{C}_1 = \hat{C}_1(T)$,

$$(6.37) \qquad \|\Delta v(t)\|_n \le \hat{C}_1\epsilon|\ln\epsilon|^2.$$

Q.E.D.

By Lemma 6.4 and estimate (6.28),

$$(6.38) \qquad \text{distance}\left\{q_\epsilon(T + \frac{1}{\mu}|\ln\epsilon|), \Pi\right\} < \tilde{C}\epsilon|\ln\epsilon|^2.$$

By Theorem 6.3, the height of the wall $W_n^s(Q_\epsilon)$ off Π is larger than the distance between $q_\epsilon(T + \frac{1}{\mu}|\ln\epsilon|)$ and Π. Thus, if $q_\epsilon(T + \frac{1}{\mu}|\ln\epsilon|)$ can move from one side of the wall $W_n^s(Q_\epsilon)$ to the other side, then by continuity $q_\epsilon(T + \frac{1}{\mu}|\ln\epsilon|)$ has to be on the wall $W_n^s(Q_\epsilon)$ at some values of the parameters.

6.4.3. Another Signed Distance. Recall the fish-like singular level set given by $\mathcal{H}$ (6.22), the width of the fish is of order $\mathcal{O}(\sqrt{\epsilon})$, and the length of the fish is of order $\mathcal{O}(1)$. Notice also that $q_0(t)$ has a phase shift

$$(6.39) \qquad \theta_1^0 = \theta^0(T + \frac{1}{\mu}|\ln\epsilon|) - \theta^0(0).$$

For fixed β, changing α can induce $\mathcal{O}(1)$ change in the length of the fish, $\mathcal{O}(\sqrt{\epsilon})$ change in θ_1^0, and $\mathcal{O}(1)$ change in $\theta^0(0)$. See Figure 5 for an illustration. The leading order signed distance from $q_\epsilon(T + \frac{1}{\mu}|\ln\epsilon|)$ to $W_n^s(Q_\epsilon)$ can be defined as

$$(6.40) \qquad \begin{aligned} \tilde{d} &= \mathcal{H}(j_0, \theta^0(0)) - \mathcal{H}(j_0, \theta^0(0) + \theta_1^0) \\ &= 2\omega\left[\alpha\omega\theta_1^0 + \beta[\sin\theta^0(0) - \sin(\theta^0(0) + \theta_1^0)]\right], \end{aligned}$$

where $\mathcal{H}$ is given in (6.22). The common zero of M_1 (6.27) and $\tilde{d}$ and the implicit function theorem imply the existence of a homoclinic orbit asymptotic to Q_ϵ. The common roots to M_1 and $\tilde{d}$ are given by $\alpha = 1/\kappa(\omega)$, where $\kappa(\omega)$ is plotted in Figure 1.

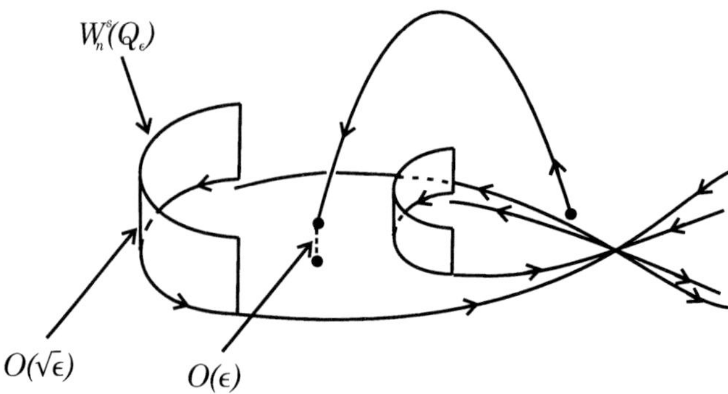

FIGURE 5. The second measurement.

6.5. Silnikov Homoclinic Orbits in Vector NLS Under Perturbations

In recent years, novel results have been obtained on the solutions of the vector nonlinear Schrödinger equations [3] [4] [206]. Abundant ordinary integrable results have been carried through [203] [70], including linear stability calculations [69]. Specifically, the vector nonlinear Schrödinger equations can be written as

$$ip_t + p_{xx} + \frac{1}{2}(|p|^2 + \chi|q|^2)p = 0,$$

$$iq_t + q_{xx} + \frac{1}{2}(\chi|p|^2 + |q|^2)q = 0,$$

where p and q are complex valued functions of the two real variables t and x, and χ is a positive constant. These equations describe the evolution of two orthogonal pulse envelopes in birefringent optical fibers [159] [160], with industrial applications in fiber communication systems [81] and all-optical switching devices [93]. For linearly birefringent fibers [159], $\chi = 2/3$. For elliptically birefringent fibers, χ can take other positive values [160]. When $\chi = 1$, these equations are first shown to be integrable by S. Manakov [150], and thus called Manakov equations. When χ is not 1 or 0, these equations are non-integrable. Propelled by the industrial applications, extensive mathematical studies on the vector nonlinear Schrödinger equations have been conducted. Like the scalar nonlinear Schrödinger equation, the vector nonlinear Schrödinger equations also possess figure eight structures in their phase space. Consider the singularly perturbed vector nonlinear Schrödinger equations,

$$(6.41) \qquad ip_t + p_{xx} + \frac{1}{2}[(|p|^2 + |q|^2) - \omega^2]p = i\epsilon[p_{xx} - \alpha p - \beta] \, ,$$

$$(6.42) \qquad iq_t + q_{xx} + \frac{1}{2}[(|p|^2 + |q|^2) - \omega^2]q = i\epsilon[q_{xx} - \alpha q - \beta] \, ,$$

where $p(t,x)$ and $q(t,x)$ are subject to periodic boundary condition of period 2π, and are even in x, i.e.

$$p(t, x + 2\pi) = p(t, x) \, , \quad p(t, -x) = p(t, x) \, ,$$

$$q(t, x + 2\pi) = q(t, x) \, , \quad q(t, -x) = q(t, x) \, ,$$

$\omega \in (1,2)$, $\alpha > 0$ and β are real constants, and $\epsilon > 0$ is the perturbation parameter. We have

THEOREM 6.5 ([**135**]). *There exists a $\epsilon_0 > 0$, such that for any $\epsilon \in (0, \epsilon_0)$, there exists a codimension 1 surface in the space of $(\alpha, \beta, \omega) \in \mathbb{R}^+ \times \mathbb{R}^+ \times \mathbb{R}^+$ where $\omega \in (1,2)/S$, S is a finite subset, and $\alpha\omega < \sqrt{2}\beta$. For any (α, β, ω) on the codimension 1 surface, the singularly perturbed vector nonlinear Schrödinger equations (6.41)-(6.42) possesses a homoclinic orbit asymptotic to a saddle Q_ϵ. This orbit is also the homoclinic orbit for the singularly perturbed scalar nonlinear Schrödinger equation studied in last section, and is the only one asymptotic to the saddle Q_ϵ for the singularly perturbed vector nonlinear Schrödinger equations (6.41)-(6.42). The codimension 1 surface has the approximate representation given by $\alpha = 1/\kappa(\omega)$, where $\kappa(\omega)$ is plotted in Figure 1.*

6.6. Silnikov Homoclinic Orbits in Discrete NLS Under Perturbations

We continue from section 5.6 and consider the perturbed discrete nonlinear Schrödinger equation (5.55). The following theorem was proved in [**138**].

Denote by Σ_N $(N \geq 7)$ the external parameter space,

$$\Sigma_N = \left\{ (\omega, \alpha, \beta) \;\middle|\; \omega \in (N \tan \frac{\pi}{N}, N \tan \frac{2\pi}{N}), \right.$$

$$\alpha \in (0, \alpha_0), \beta \in (0, \beta_0);$$

$$\left. \text{where } \alpha_0 \text{ and } \beta_0 \text{ are any fixed positive numbers.} \right\}$$

THEOREM 6.6. *For any N $(7 \leq N < \infty)$, there exists a positive number ϵ_0, such that for any $\epsilon \in (0, \epsilon_0)$, there exists a codimension 1 surface E_ϵ in Σ_N; for any external parameters (ω, α, β) on E_ϵ, there exists a homoclinic orbit asymptotic to a saddle Q_ϵ. The codimension 1 surface E_ϵ has the approximate expression $\alpha = 1/\kappa$, where $\kappa = \kappa(\omega; N)$ is shown in Fig.6.*

In the cases $(3 \leq N \leq 6)$, κ is always negative. For $N \geq 7$, κ can be positive as shown in Fig.6. When N is even and ≥ 7, there is in fact a symmetric pair of homoclinic orbits asymptotic to a fixed point Q_ϵ at the same values of the external parameters; since for even N, we have the symmetry: If $q_n = f(n, t)$ solves (5.55), then $q_n = f(n + N/2, t)$ also solves (5.55). When N is odd and ≥ 7, the study can not guarantee that two homoclinic orbits exist at the same value of the external parameters.

6.7. Comments on DSII Under Perturbations

We continue from section 5.5. Under both regular and singular perturbations, the rigorous Melnikov measurement can be established [**132**]. It turns out that only local well-posedness is necessary for rigorously setting up the Melnikov measurement, thanks to the fact that the unperturbed homoclinic orbits given in section 3.3.1 are classical solutions. Thus the Melnikov integrals given in section 4.3.2 indeed rigorously measure signed distances. For details, see [**132**].

The obstacle toward proving the existence of homoclinic orbits comes from a technical difficulty in solving a linear system to get the normal form for proving the size estimate of the stable manifold of the saddle. For details, see [**132**].

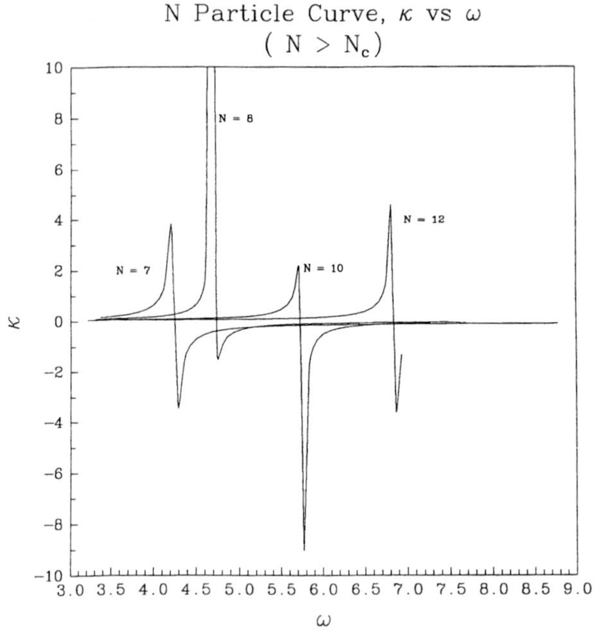

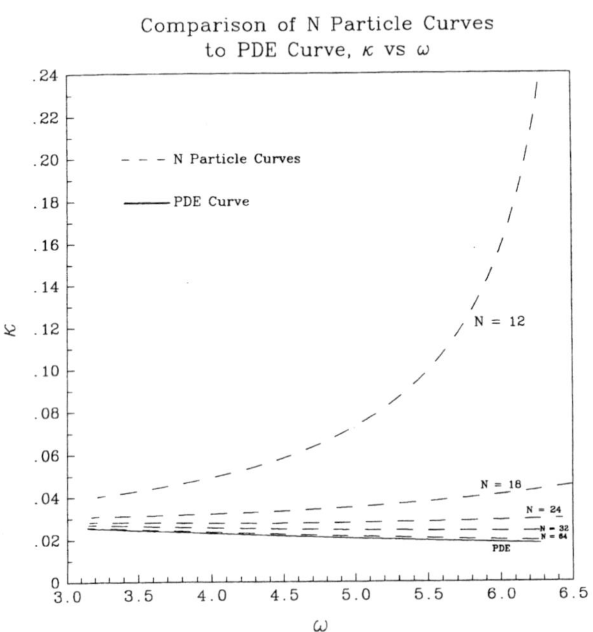

FIGURE 6. The graph of $\kappa(\omega; N)$.

6.8. Normal Form Transforms

Consider the singularly perturbed Davey-Stewartson II equation in section 5.5,

$$(6.43) \qquad \begin{cases} iq_t = \Upsilon q + [2(|q|^2 - \omega^2) + u_y]q + i\epsilon[\Delta q - \alpha q + \beta] \,, \\ \Delta u = -4\partial_y |q|^2 \,. \end{cases}$$

Let

$$q(t,x,y) = [\rho(t) + f(t,x,y)]e^{i\theta(t)}, \quad \langle f \rangle = 0,$$

where $\langle\,,\,\rangle$ denotes spatial mean. Let

$$I = \langle |q|^2 \rangle = \rho^2 + \langle |f|^2 \rangle, \quad J = I - \omega^2.$$

In terms of the new variables (J, θ, f), Equation (6.43) can be rewritten as

$$(6.44) \qquad \dot{J} = \epsilon \left[-2\alpha(J + \omega^2) + 2\beta\sqrt{J + \omega^2}\cos\theta \right] + \epsilon \widehat{R}_2^J,$$

$$(6.45) \qquad \dot{\theta} = -2J - \epsilon\beta\frac{\sin\theta}{\sqrt{J + \omega^2}} + \widehat{R}_2^\theta,$$

$$(6.46) \qquad f_t = L_\epsilon f + V_\epsilon f - i\mathcal{N}_2 - i\mathcal{N}_3,$$

where

$$L_\epsilon f = -i\Upsilon f + \epsilon(\Delta - \alpha)f - 2i\omega^2\Delta^{-1}\Upsilon(f + \bar{f}),$$

$$\mathcal{N}_2 = 2\omega\left[\Delta^{-1}\Upsilon|f|^2 + f\Delta^{-1}\Upsilon(f + \bar{f}) - \langle f\Delta^{-1}\Upsilon(f + \bar{f})\rangle \right],$$

and $\widehat{R}_2^J$, $\widehat{R}_2^\theta$, $V_\epsilon f$, and $\mathcal{N}_3$ are higher order terms. It is the quadratic term $\mathcal{N}_2$ that blocks the size estimate of the stable manifold of the saddle [**132**]. Thus, our goal is to find a normal form transform $g = f + K(f,f)$ where K is a bilinear form, that transforms the equation

$$f_t = L_\epsilon f - i\mathcal{N}_2,$$

into an equation with a cubic nonlinearity

$$g_t = L_\epsilon g + \mathcal{O}(\|g\|_s^3), \quad (s \geq 2),$$

where L_ϵ is given in (6.46). In terms of Fourier transforms,

$$f = \sum_{k \neq 0} \hat{f}(k)e^{ik\cdot\xi}, \quad \bar{f} = \sum_{k \neq 0} \overline{\hat{f}(-k)}e^{ik\cdot\xi},$$

where $k = (k_1, k_2) \in \mathbb{Z}^2$, $\xi = (\kappa_1 x, \kappa_2 y)$. The terms in $\mathcal{N}_2$ can be written as

$$\Delta^{-1}\Upsilon|f|^2 = \frac{1}{2}\sum_{k+\ell \neq 0} a(k + \ell)\left[\hat{f}(k)\overline{\hat{f}(-\ell)} + \hat{f}(\ell)\overline{\hat{f}(-k)} \right]e^{i(k+\ell)\cdot\xi},$$

$$f\Delta^{-1}\Upsilon f - \langle f\Delta^{-1}\Upsilon f\rangle = \frac{1}{2}\sum_{k+\ell \neq 0} [a(k) + a(\ell)]\hat{f}(k)\hat{f}(\ell)e^{i(k+\ell)\cdot\xi},$$

$$f\Delta^{-1}\Upsilon\bar{f} - \langle f\Delta^{-1}\Upsilon\bar{f}\rangle = \frac{1}{2}\sum_{k+\ell \neq 0} \left[a(\ell)\hat{f}(k)\overline{\hat{f}(-\ell)} + a(k)\hat{f}(\ell)\overline{\hat{f}(-k)} \right]e^{i(k+\ell)\cdot\xi},$$

where

$$a(k) = \frac{k_1^2\kappa_1^2 - k_2^2\kappa_2^2}{k_1^2\kappa_1^2 + k_2^2\kappa_2^2}.$$

We will search for a normal form transform of the general form,

$$g = f + K(f,f),$$

where

$$K(f, f) = \sum_{k+\ell \neq 0} \left[\hat{K}_1(k, \ell) \hat{f}(k) \hat{f}(\ell) + \hat{K}_2(k, \ell) \hat{f}(k) \overline{\hat{f}(-\ell)} \right.$$

$$\left. + \hat{K}_2(\ell, k) \overline{\hat{f}(-k)} \hat{f}(\ell) + \hat{K}_3(k, \ell) \overline{\hat{f}(-k)} \overline{\hat{f}(-\ell)} \right] e^{i(k+\ell)x},$$

where $\hat{K}_j(k, \ell)$, $(j = 1, 2, 3)$ are the unknown coefficients to be determined, and $\hat{K}_j(k, \ell) = \hat{K}_j(\ell, k)$, $(j = 1, 3)$. To eliminate the quadratic terms, we first need to set

$$iL_\epsilon K(f, f) - iK(L_\epsilon f, f) - iK(f, L_\epsilon f) = \mathcal{N}_2,$$

which takes the explicit form:

$$(\sigma_1 + i\sigma) \hat{K}_1(k, \ell) + B(\ell) \hat{K}_2(k, \ell) + B(k) \hat{K}_2(\ell, k)$$

$$(6.47) \qquad + B(k + \ell) \overline{\hat{K}_3(k, \ell)} = \frac{1}{2\omega} [B(k) + B(\ell)],$$

$$-B(\ell) \hat{K}_1(k, \ell) + (\sigma_2 + i\sigma) \hat{K}_2(k, \ell) + B(k + \ell) \overline{\hat{K}_2(\ell, k)}$$

$$(6.48) \qquad + B(k) \hat{K}_3(k, \ell) = \frac{1}{2\omega} [B(k + \ell) + B(\ell)],$$

$$-B(k) \hat{K}_1(k, \ell) + B(k + \ell) \overline{\hat{K}_2(k, \ell)} + (\sigma_3 + i\sigma) \hat{K}_2(\ell, k)$$

$$(6.49) \qquad + B(\ell) \hat{K}_3(k, \ell) = \frac{1}{2\omega} [B(k + \ell) + B(k)],$$

$$B(k + \ell) \overline{\hat{K}_1(k, \ell)} - B(k) \hat{K}_2(k, \ell) - B(\ell) \hat{K}_2(\ell, k)$$

$$(6.50) \qquad + (\sigma_4 + i\sigma) \hat{K}_3(k, \ell) = 0,$$

where $B(k) = 2\omega^2 a(k)$, and

$$\sigma = \epsilon \left[\alpha - 2(k_1 \ell_1 \kappa_1^2 + k_2 \ell_2 \kappa_2^2) \right],$$

$$\sigma_1 = 2(k_2 \ell_2 \kappa_2^2 - k_1 \ell_1 \kappa_1^2) + B(k + \ell) - B(k) - B(\ell),$$

$$\sigma_2 = 2[(k_2 + \ell_2)\ell_2 \kappa_2^2 - (k_1 + \ell_1)\ell_1 \kappa_1^2] + B(k + \ell) - B(k) + B(\ell),$$

$$\sigma_3 = 2[(k_2 + \ell_2)k_2 \kappa_2^2 - (k_1 + \ell_1)k_1 \kappa_1^2] + B(k + \ell) + B(k) - B(\ell),$$

$$\sigma_4 = 2[(k_2^2 + k_2 \ell_2 + \ell_2^2)\kappa_2^2 - (k_1^2 + k_1 \ell_1 + \ell_1^2)\kappa_1^2] + B(k + \ell) + B(k) + B(\ell).$$

Since these coefficients are even in (k, ℓ), we will search for even solutions, i.e.

$$\hat{K}_j(-k, -\ell) = \hat{K}_j(k, \ell), \quad j = 1, 2, 3.$$

The technical difficulty in the normal form transform comes from not being able to answer the following two questions in solving the linear system (6.47)-(6.50):

(1) Is it true that for all $k, \ell \in \mathbb{Z}^2/\{0\}$, there is a solution ?
(2) What is the asymptotic behavior of the solution as k and/or $\ell \to \infty$? In particular, is the asymptotic behavior like k^{-m} and/or ℓ^{-m} $(m \geq 0)$?

Setting $\partial_y = 0$ in (6.43), the singularly perturbed Davey-Stewartson II equation reduces to the singularly perturbed nonlinear Schrödinger (NLS) equation (5.2) in section 5.2. Then the above two questions can be answered [124]. For the regularly perturbed nonlinear Schrödinger (NLS) equation (5.1) in section 5.1, the answer to the above two questions is even simpler [136] [124].

6.9. Transversal Homoclinic Orbits in a Periodically Perturbed SG

Transversal homoclinic orbits in continuous systems often appear in two types of systems: (1). periodic systems where the Poincaré period map has a transversal homoclinic orbit, (2). autonomous systems where the homoclinic orbit is asymptotic to a hyperbolic limit cycle.

Consider the periodically perturbed sine-Gordon (SG) equation,

$$(6.51) \qquad u_{tt} = c^2 u_{xx} + \sin u + \epsilon[-au_t + u^3 \chi(\|u\|) \cos t],$$

where

$$\chi(\|u\|) = \left\{ \begin{array}{ll} 1, & \|u\| \le M, \\ 0, & \|u\| \ge 2M, \end{array} \right.$$

for $M < \|u\| < 2M$, $\chi(\|u\|)$ is a smooth bump function, under odd periodic boundary condition,

$$u(x + 2\pi, t) = u(x, t), \quad u(x, t) = -u(x, t),$$

$\frac{1}{4} < c^2 < 1$, $a > 0$, ϵ is a small perturbation parameter.

THEOREM 6.7 ([**136**], [**183**]). *There exists an interval $I \subset \mathbb{R}^+$ such that for any $a \in I$, there exists a transversal homoclinic orbit $u = \xi(x, t)$ asymptotic to 0 in H^1.*

6.10. Transversal Homoclinic Orbits in a Derivative NLS

Consider the derivative nonlinear Schrödinger equation,

$$(6.52) \qquad iq_t = q_{xx} + 2|q|^2 q + i\epsilon\left[(\frac{9}{16} - |q|^2)q + \mu|\hat{\partial}_x q|^2 \bar{q}\right],$$

where q is a complex-valued function of two real variables t and x, $\epsilon > 0$ is the perturbation parameter, μ is a real constant, and $\hat{\partial}_x$ is a bounded Fourier multiplier,

$$\hat{\partial}_x q = -\sum_{k=1}^{K} k\tilde{q}_k \sin kx , \quad \text{for } q = \sum_{k=0}^{\infty} \tilde{q}_k \cos kx ,$$

and some fixed K. Periodic boundary condition and even constraint are imposed,

$$q(t, x + 2\pi) = q(t, x) , \quad q(t, -x) = q(t, x) .$$

THEOREM 6.8 ([**125**]). *There exists a $\epsilon_0 > 0$, such that for any $\epsilon \in (0, \epsilon_0)$, and $|\mu| > 5.8$, there exist two transversal homoclinic orbits asymptotic to the limit cycle $q_c = \frac{3}{4} \exp\{-i[\frac{9}{8}t + \gamma]\}$.*

CHAPTER 7

Existence of Chaos

The importance of homoclinic orbits with respect to chaotic dynamics was first realized by Poincaré [177]. In 1961, Smale constructed the well-known horseshoe in the neighborhood of a transversal homoclinic orbit [190] [191] [192]. In particular, Smale's theorem implies Birkhoff's theorem on the existence of a sequence of structurely stable periodic orbits in the neighborhood of a transversal homoclinic orbit [27]. In 1984 and 1988 [172] [173], Palmer gave a beautiful proof of Smale's theorem using a shadowing lemma. Later, this proof was generalized to infinite dimensions by Steinlein and Walther [193] [194] and Henry [83]. In 1967, Silnikov proved Smale's theorem for autonomous systems in finite dimensions using a fixed point argument [186]. In 1996, Palmer proved Smale's theorem for autonomous systems in finite dimensions using shadowing lemma [50] [174]. In 2002, Li proved Smale's theorem for autonomous systems in infinite dimensions using shadowing lemma [125].

For nontransversal homoclinic orbits, the most well-known type which leads to the existence of Smale horseshoes is the so-called Silnikov homoclinic orbit [184] [185] [188] [55] [56]. Existence of Silnikov homoclinic orbits and new constructions of Smale horseshoes for concrete nonlinear wave systems have been established in finite dimensions [138] [139] and in infinite dimensions [136] [120] [124].

7.1. Horseshoes and Chaos

7.1.1. Equivariant Smooth Linearization. Linearization has been a popular topics in dynamical systems. It asks the question whether or not one can transform a nonlinear system into its linearized system at a fixed point, in a small neighborhood of the fixed point. For a sample of references, see [80] [182]. Here we specifically consider the singularly perturbed nonlinear Schrödinger (NLS) equation (5.2) in sections 5.2 and 6.2. The symmetric pair of Silnikov homoclinic orbits is asymptotic to the saddle $Q_\epsilon = \sqrt{I} e^{i\theta}$, where

$$(7.1) \qquad I = \omega^2 - \epsilon \frac{1}{2\omega} \sqrt{\beta^2 - \alpha^2 \omega^2} + \cdots, \qquad \cos\theta = \frac{\alpha\sqrt{I}}{\beta}, \qquad \theta \in (0, \frac{\pi}{2}).$$

Its eigenvalues are

$$(7.2) \qquad \lambda_n^\pm = -\epsilon[\alpha + n^2] \pm 2\sqrt{(\frac{n^2}{2} + \omega^2 - I)(3I - \omega^2 - \frac{n^2}{2})} \,,$$

where $n = 0, 1, 2, \cdots$, $\omega \in (\frac{1}{2}, 1)$, and I is given in (7.1). In conducting linearization, the crucial factor is the so-callled nonresonance conditions.

LEMMA 7.1 ([120] [129]). *For any fixed $\epsilon \in (0, \epsilon_0)$, let E_ϵ be the codimension 1 surface in the external parameter space, on which the symmetric pair of Silnikov homoclinic orbits are supported (cf: Theorem 6.2). For almost every $(\alpha, \beta, \omega) \in E_\epsilon$,*

*the eigenvalues $\lambda_n^\pm$ (6.2) satisfy the nonresonance condition of Siegel type: There
exists a natural number s such that for any integer $n \geq 2$,*

$$\left| \Lambda_n - \sum_{j=1}^{r} \Lambda_{l_j} \right| \geq 1/r^s \ ,$$

*for all $r = 2, 3, \cdots, n$ and all $l_1, l_2, \cdots, l_r \in \mathbb{Z}$, where $\Lambda_n = \lambda_n^+$ for $n \geq 0$, and
$\Lambda_n = \lambda_{-n-1}^-$ for $n < 0$.*

Thus, in a neighborhood of Q_ϵ, the singularly perturbed NLS (5.2) is analyti-
cally equivalent to its linearization at Q_ϵ [**166**]. In terms of eigenvector basis, (5.2)
can be rewritten as

$$
\begin{aligned}
\dot{x} &= -ax - by + \mathcal{N}_x(\vec{Q}), \\
\dot{y} &= bx - ay + \mathcal{N}_y(\vec{Q}), \\
\dot{z}_1 &= \gamma_1 z_1 + \mathcal{N}_{z_1}(\vec{Q}), \\
\dot{z}_2 &= \gamma_2 z_2 + \mathcal{N}_{z_2}(\vec{Q}), \\
\dot{Q} &= LQ + \mathcal{N}_Q(\vec{Q});
\end{aligned}
$$

(7.3)

where $a = -\mathrm{Re}\{\lambda_2^+\}$, $b = \mathrm{Im}\{\lambda_2^+\}$, $\gamma_1 = \lambda_0^+$, $\gamma_2 = \lambda_1^+$; $\mathcal{N}$'s vanish identically in
a neighborhood Ω of $\vec{Q} = 0$, $\vec{Q} = (x, y, z_1, z_2, Q)$, Q is associated with the rest of
eigenvalues, L is given as

$$LQ = -iQ_{\zeta\zeta} - 2i[(2|Q_\epsilon|^2 - \omega^2)Q + Q_\epsilon^2 \bar{Q}] + \epsilon[-\alpha Q + Q_{\zeta\zeta}] \ ,$$

and Q_ϵ is given in (7.1).

7.1.2. Conley-Moser Conditions. We continue from last section. Denote
by h_k ($k = 1, 2$) the symmetric pair of Silnikov homoclinic orbits. The symmetry
σ of half spatial period shifting has the new representation in terms of the new
coordinates

(7.4) $$\sigma \circ (x, y, z_1, z_2, Q) = (x, y, z_1, -z_2, \sigma \circ Q).$$

DEFINITION 7.2. The Poincaré section Σ_0 is defined by the constraints:

$$
\begin{aligned}
y &= 0, \ \eta \exp\{-2\pi a/b\} < x < \eta; \\
0 &< z_1 < \eta, \ -\eta < z_2 < \eta, \ \|Q\| < \eta;
\end{aligned}
$$

where η is a small parameter.

The horseshoes are going to be constructed on this Poincaré section.

DEFINITION 7.3. The auxiliary Poincaré section Σ_1 is defined by the constraints:

$$
\begin{aligned}
z_1 &= \eta, \quad -\eta < z_2 < \eta, \\
\sqrt{x^2 + y^2} &< \eta, \quad \|Q\| < \eta.
\end{aligned}
$$

Maps from a Poincaré section to a Poincaré section are induced by the flow.
Let $\vec{Q}^0$ and $\vec{Q}^1$ be the coordinates on Σ_0 and Σ_1 respectively, then the map P_0^1

from Σ_0 to Σ_1 has the simple expression:

$$
\begin{aligned}
x^1 &= \left(\frac{z_1^0}{\eta}\right)^{\frac{a}{\gamma_1}} x^0 \cos\left[\frac{b}{\gamma_1} \ln \frac{\eta}{z_1^0}\right] , \\
y^1 &= \left(\frac{z_1^0}{\eta}\right)^{\frac{a}{\gamma_1}} x^0 \sin\left[\frac{b}{\gamma_1} \ln \frac{\eta}{z_1^0}\right] , \\
z_2^1 &= \left(\frac{\eta}{z_1^0}\right)^{\frac{\gamma_2}{\gamma_1}} z_2^0 , \\
Q^1 &= e^{t_0 L} Q^0 .
\end{aligned}
$$

Let $\vec{Q}_*^0$ and $\vec{Q}_*^1$ be the intersection points of the homoclinic orbit h_1 with $\overline{\Sigma_0}$ and Σ_1 respectively, and let $\vec{\tilde{Q}}^0 = \vec{Q}^0 - \vec{Q}_*^0$, and $\vec{\tilde{Q}}^1 = \vec{Q}^1 - \vec{Q}_*^1$. One can define slabs in Σ_0.

DEFINITION 7.4. For sufficiently large natural number l, we define slab S_l in Σ_0 as follows:

$$
\begin{aligned}
S_l \equiv \Big\{ \vec{Q} \in \Sigma_0 \ \Big| \ & \eta \exp\{-\gamma_1(t_{0,2(l+1)} - \frac{\pi}{2b})\} \le \\
& \tilde{z}_1^0(\vec{Q}) \le \eta \exp\{-\gamma_1(t_{0,2l} - \frac{\pi}{2b})\}, \\
& |\tilde{x}^0(\vec{Q})| \le \eta \exp\{-\frac{1}{2}a\, t_{0,2l}\}, \\
& |\tilde{z}_2^1(P_0^1(\vec{Q}))| \le \eta \exp\{-\frac{1}{2}a\, t_{0,2l}\}, \\
& \|\tilde{Q}^1(P_0^1(\vec{Q}))\| \le \eta \exp\{-\frac{1}{2}a\, t_{0,2l}\} \Big\},
\end{aligned}
$$

where

$$
t_{0,l} = \frac{1}{b}[l\pi - \varphi_1] + o(1) ,
$$

as $l \to +\infty$, are the time that label the fixed points of the Poincaré map P from Σ_0 to Σ_0 [120] [129], and the notations $\tilde{x}^0(\vec{Q})$, $\tilde{z}_2^1(P_0^1(\vec{Q}))$, etc. denote the $\tilde{x}^0$ coordinate of the point $\vec{Q}$, the $\tilde{z}_2^1$ coordinate of the point $P_0^1(\vec{Q})$, etc..

S_l is defined so that it includes two fixed points p_l^+ and p_l^- of P. Let $S_{l,\sigma} = \sigma \circ S_l$ where the symmetry σ is defined in (7.4). We need to define a larger slab $\hat{S}_l$ such that $S_l \cup S_{l,\sigma} \subset \hat{S}_l$.

DEFINITION 7.5. The larger slab $\hat{S}_l$ is defined as

$$
\begin{aligned}
\hat{S}_l = \Big\{ \vec{Q} \in \Sigma_0 \ \Big| \ & \eta \exp\{-\gamma_1(t_{0,2(l+1)} - \frac{\pi}{2b})\} \le \\
& z_1^0(\vec{Q}) \le \eta \exp\{-\gamma_1(t_{0,2l} - \frac{\pi}{2b})\}, \\
& |x^0(\vec{Q}) - x_*^0| \le \eta \exp\{-\frac{1}{2}a\, t_{0,2l}\}, \\
& |z_2^1(P_0^1(\vec{Q}))| \le |z_{2,*}^1| + \eta \exp\{-\frac{1}{2}a\, t_{0,2l}\}, \\
& \|Q^1(P_0^1(\vec{Q}))\| \le \eta \exp\{-\frac{1}{2}a\, t_{0,2l}\} \Big\},
\end{aligned}
$$

where $z_{2,*}^1$ is the z_2^1-coordinate of $\vec{Q}_*^1$.

DEFINITION 7.6. In the coordinate system $\{\tilde{x}^0, \tilde{z}_1^0, \tilde{z}_2^0, \tilde{Q}^0\}$, the stable boundary of $\hat{S}_l$, denoted by $\partial_s \hat{S}_l$, is defined to be the boundary of $\hat{S}_l$ along $(\tilde{x}^0, \tilde{Q}^0)$-directions, and the unstable boundary of $\hat{S}_l$, denoted by $\partial_u \hat{S}_l$, is defined to be the boundary of $\hat{S}_l$ along $(\tilde{z}_1^0, \tilde{z}_2^0)$-directions. A stable slice V in $\hat{S}_l$ is a subset of $\hat{S}_l$, defined as the region swept out through homeomorphically moving and deforming $\partial_s \hat{S}_l$ in such a way that the part

$$\partial_s \hat{S}_l \cap \partial_u \hat{S}_l$$

of $\partial_s \hat{S}_l$ only moves and deforms inside $\partial_u \hat{S}_l$. The new boundary obtained through such moving and deforming of $\partial_s \hat{S}_l$ is called the stable boundary of V, which is denoted by $\partial_s V$. The rest of the boundary of V is called its unstable boundary, which is denoted by $\partial_u V$. An unstable slice of $\hat{S}_l$, denoted by H, is defined similarly.

As shown in [120] and [129], under certain generic assumptions, when l is sufficiently large, $P(S_l)$ and $P(S_{l,\sigma})$ intersect $\hat{S}_l$ into four disjoint stable slices $\{V_1, V_2\}$ and $\{V_{-1}, V_{-2}\}$ in $\hat{S}_l$. V_j's $(j = 1, 2, -1, -2)$ do not intersect $\partial_s \hat{S}_l$; moreover,

$$(7.5) \qquad \partial_s V_i \subset P(\partial_s S_l), (i = 1, 2); \quad \partial_s V_i \subset P(\partial_s S_{l,\sigma}), (i = -1, -2).$$

Let

$$(7.6) \qquad H_j = P^{-1}(V_j), \quad (j = 1, 2, -1, -2),$$

where and for the rest of this article, P^{-1} denotes preimage of P. Then H_j $(j = 1, 2, -1, -2)$ are unstable slices. More importantly, the Conley-Moser conditions are satisfied. Specifically, Conley-Moser conditions are: $\boxed{\text{Conley-Moser condition (i):}}$

$$\begin{cases} V_j = P(H_j), \\ \partial_s V_j = P(\partial_s H_j), & (j = 1, 2, -1, -2) \\ \partial_u V_j = P(\partial_u H_j). \end{cases}$$

$\boxed{\text{Conley-Moser condition (ii):}}$ There exists a constant $0 < \nu < 1$, such that for any stable slice $V \subset V_j$ $(j = 1, 2, -1, -2)$, the diameter decay relation

$$d(\tilde{V}) \leq \nu d(V)$$

holds, where $d(\cdot)$ denotes the diameter [120], and $\tilde{V} = P(V \cap H_k)$, $(k = 1, 2, -1, -2)$; for any unstable slice $H \subset H_j$ $(j = 1, 2, -1, -2)$, the diameter decay relation

$$d(\tilde{H}) \leq \nu d(H)$$

holds, where $\tilde{H} = P^{-1}(H \cap V_k)$, $(k = 1, 2, -1, -2)$.

The Conley-Moser conditions are sufficient conditions for establishing the topological conjugacy between the Poincare map P restricted to a Cantor set in Σ_0, and the shift automorphism on symbols. It also display the horseshoe nature of the intersection of S_l and $S_{l,\sigma}$ with their images $P(S_l)$ and $P(S_{l,\sigma})$.

7.1.3. Shift Automorphism. Let $\mathcal{W}$ be a set which consists of elements of the doubly infinite sequence form:

$$a = (\cdots a_{-2} a_{-1} a_0, a_1 a_2 \cdots),$$

where $a_k \in \{1, 2, -1, -2\}$; $k \in \mathbb{Z}$. We introduce a topology in $\mathcal{W}$ by taking as neighborhood basis of

$$a^* = (\cdots a_{-2}^* a_{-1}^* a_0^*, a_1^* a_2^* \cdots),$$

the set

$$W_j = \left\{ a \in \mathcal{W} \;\middle|\; a_k = a_k^* \; (|k| < j) \right\}$$

for $j = 1, 2, \cdots$. This makes $\mathcal{W}$ a topological space. The shift automorphism χ is defined on $\mathcal{W}$ by

$$\chi \;:\; \mathcal{W} \mapsto \mathcal{W},$$
$$\forall a \in \mathcal{W}, \; \chi(a) = b, \; \text{where } b_k = a_{k+1}.$$

The shift automorphism χ exhibits *sensitive dependence on initial conditions*, which is a hallmark of *chaos*.

Let

$$a = (\cdots a_{-2} a_{-1} a_0, a_1 a_2 \cdots),$$

be any element of $\mathcal{W}$. Define inductively for $k \geq 2$ the stable slices

$$V_{a_0 a_{-1}} = P(H_{a_{-1}}) \cap H_{a_0},$$
$$V_{a_0 a_{-1} \ldots a_{-k}} = P(V_{a_{-1} \ldots a_{-k}}) \cap H_{a_0}.$$

By Conley-Moser condition (ii),

$$d(V_{a_0 a_{-1} \ldots a_{-k}}) \leq \nu_1 d(V_{a_0 a_{-1} \ldots a_{-(k-1)}}) \leq \cdots \leq \nu_1^{k-1} d(V_{a_0 a_{-1}}).$$

Then,

$$V(a) = \bigcap_{k=1}^{\infty} V_{a_0 a_{-1} \ldots a_{-k}}$$

defines a 2 dimensional continuous surface in Σ_0; moreover,

$$(7.7) \qquad \partial V(a) \subset \partial_u \hat{S}_l.$$

Similarly, define inductively for $k \geq 1$ the unstable slices

$$H_{a_0 a_1} = P^{-1}(H_{a_1} \cap V_{a_0}),$$
$$H_{a_0 a_1 \ldots a_k} = P^{-1}(H_{a_1 \ldots a_k} \cap V_{a_0}).$$

By Conley-Moser condition (ii),

$$d(H_{a_0 a_1 \ldots a_k}) \leq \nu_2 d(H_{a_0 a_1 \ldots a_{k-1}}) \leq \cdots \leq \nu_2^k d(H_{a_0}).$$

Then,

$$H(a) = \bigcap_{k=0}^{\infty} H_{a_0 a_1 \ldots a_k}$$

defines a codimension 2 continuous surface in Σ_0; moreover,

$$(7.8) \qquad \partial H(a) \subset \partial_s \hat{S}_l.$$

By (7.7;7.8) and dimension count,

$$V(a) \cap H(a) \neq \emptyset$$

consists of points. Let

$$p \in V(a) \cap H(a)$$

be any point in the intersection set. Now we define the mapping

$$\phi \;:\; \mathcal{W} \mapsto \hat{S}_l,$$
$$\phi(a) = p.$$

By the above construction,

$$P(p) = \phi(\chi(a)).$$

That is,

$$P \circ \phi = \phi \circ \chi.$$

Let

$$\Lambda \equiv \phi(\mathcal{W}),$$

then Λ is a compact Cantor subset of $\hat{S}_l$, and invariant under the Poincare map P. Moreover, with the topology inherited from $\hat{S}_l$ for Λ, ϕ is a homeomorphism from $\mathcal{W}$ to Λ.

7.2. Nonlinear Schrödinger Equation Under Singular Perturbations

Finally, one has the theorem on the existence of chaos in the singularly perturbed nonlinear Schrödinger system (5.2).

THEOREM 7.7 (Chaos Theorem, [**129**]). *Under certain generic assumptions, for the perturbed nonlinear Schrödinger system (5.2), there exists a compact Cantor subset Λ of $\hat{S}_l$, Λ consists of points, and is invariant under P. P restricted to Λ, is topologically conjugate to the shift automorphism χ on four symbols $1, 2, -1, -2$. That is, there exists a homeomorphism*

$$\phi \; : \; \mathcal{W} \mapsto \Lambda,$$

such that the following diagram commutes:

$$
\begin{array}{ccc}
\mathcal{W} & \xrightarrow{\;\phi\;} & \Lambda \\
\chi \downarrow & & \downarrow P \\
\mathcal{W} & \xrightarrow[\phi]{} & \Lambda
\end{array}
$$

Although the symmetric pair of Silnikov homoclinic orbits is not structurally stable, the Smale horseshoes are structurally stable. Thus, the Cantor sets and the conjugacy to the shift automorphism are structurally stable.

7.3. Nonlinear Schrödinger Equation Under Regular Perturbations

We continue from section 6.1 and consider the regularly perturbed nonlinear Schrödinger (NLS) equation (5.1). Starting from the Homoclinic Orbit Theorem 6.1, the arguments in previous sections equally apply. In fact, under the regular perturbation, the argument can be simplified due to the fact that the evolution operator is a group. Thus, the Chaos Theorem 7.7 holds for the regularly perturbed nonlinear Schrödinger (NLS) equation (5.1).

7.4. Discrete Nonlinear Schrödinger Equation Under Perturbations

We continue from section 6.6. Starting from the Homoclinic Orbit Theorem 6.6, the arguments in previous sections equally apply and are easier. Thus, the Chaos Theorem 7.7 holds for the perturbed discrete nonlinear Schrödinger equation (5.55). In the case $N \geq 7$ and is odd, we only define one slab S_l. Consequently, the Poincaré map is topologically conjugate to the shift automorphism on two symbols.

7.5. Numerical Simulation of Chaos

The finite-difference discretization of both the regularly and the singularly perturbed nonlinear Schrödinger equations (5.1) and (5.2) leads to the same discrete perturbed nonlinear Schrödinger equation (5.55).

In the chaotic regime, typical numerical output is shown in Figure 1. Notice that there are two typical profiles at a fixed time: (1). a breather type profile with its hump located at the center of the spatial period interval, (2). a breather type profile with its hump located at the boundary (wing) of the spatial period interval. These two types of profiles are half spatial period translate of each other. If we label the profiles, with their humps at the center of the spatial period interval, by "C"; those profiles, with their humps at the wing of the spatial period interval, by "W", then

$$(7.10) \qquad\qquad "W" = \sigma \circ "C",$$

where σ is the symmetry group element representing half spatial period translate. The time series of the output in Figure 1 is a chaotic jumping between "C" and "W", which we call "chaotic center-wing jumping". By the relation (7.10) and the study in previous sections, we interpret the chaotic center-wing jumping as the numerical realization of the shift automorphism χ on four symbols $1, 2, -1, -2$. We can make this more precise in terms of the phase space geometry. The figure-eight structure (3.14) projected onto the plane of the Fourier component $\cos x$ is illustrated in Figure 2, and labeled by L_C and L_W. From the symmetry, we know that

$$L_W = \sigma \circ L_C.$$

L_C has the spatial-temporal profile realization as in Figure 3 with the hump located at the center. L_W corresponds to the half spatial period translate of the spatial-temporal profile realization as in Figure 3, with the hump located at the boundary (wing). An orbit inside L_C, L_{Cin} has a spatial-temporal profile realization as in Figure 4. The half period translate of L_{Cin}, L_{Win} is inside L_W. An orbit outside L_C and L_W, L_{out} has the spatial-temporal profile realization as in Figure 5. The two slabs (unstable slices of $\hat{S}_l$) S_l and $S_{l,\sigma}$ projected onto the plane of "$\cos x$", are also illustrated in Figure 2. From these figures, we can see clearly that the chaotic center-wing jumping (Figure 1) is the realization of the shift automorphism on four symbols $1, 2, -1, -2$ in $S_l \cup S_{l,\sigma}$.

7.6. Shadowing Lemma and Chaos in Finite-D Periodic Systems

Since its invention [9], shadowing lemma has been a useful tool for solving many dynamical system problems. Here we only focus upon its use in proving the existence of chaos in a neighborhood of a transversal homoclinic orbit. According to the type of the system, the level of difficulty in proving the existence of chaos with a shadowing lemma is different.

A finite-dimensional periodic system can be written in the general form

$$\dot{x} = F(x, t) ,$$

where $x \in \mathbb{R}^n$, and $F(x,t)$ is periodic in t. Let f be the Poincaré period map.

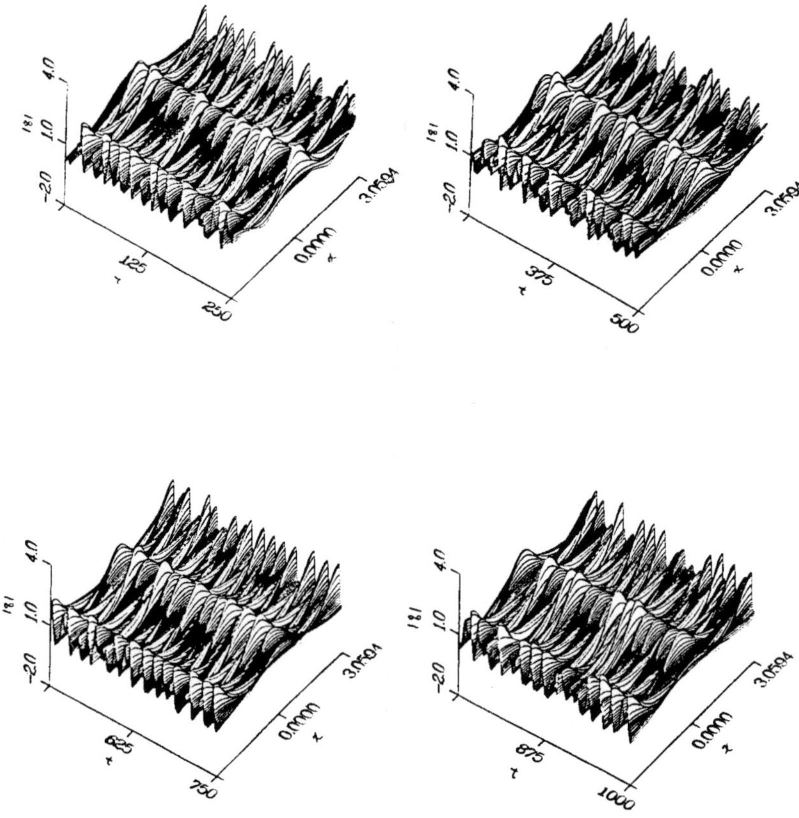

FIGURE 1. A chaotic solution in the discrete perturbed NLS system (5.55).

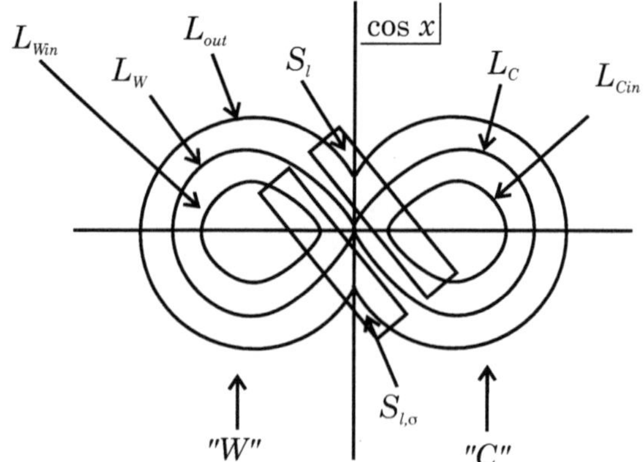

FIGURE 2. Figure-eight structure and its correspondence with the chaotic center-wing jumping.

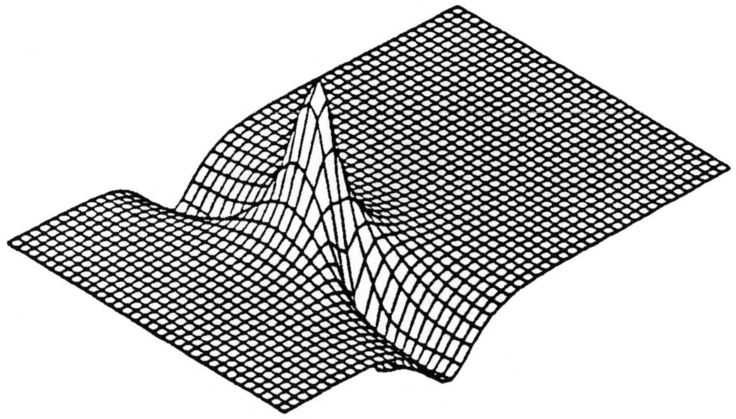

FIGURE 3. Spatial-temporal profile realization of L_C in Figure 2 (coordinates are the same as in Figure 1).

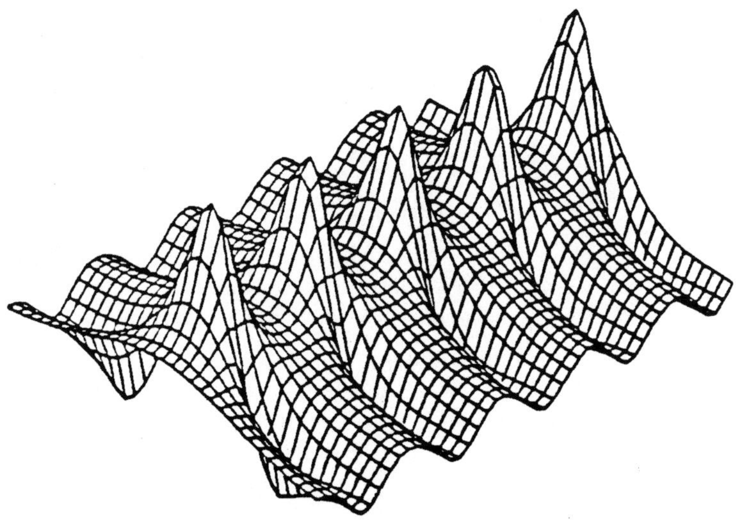

FIGURE 4. Spatial-temporal profile realization of L_{Cin} in Figure 2 (coordinates are the same as in Figure 1).

DEFINITION 7.8. A doubly infinite sequence $\{y_j\}$ in $\mathbb{R}^n$ is a δ pseudo-orbit of a C^1 diffeomorphism $f : \mathbb{R}^n \mapsto \mathbb{R}^n$ if for all integers j

$$|y_{j+1} - f(y_j)| \leq \delta .$$

An orbit $\{f^j(x)\}$ is said to ϵ-shadow the δ pseudo-orbit $\{y_j\}$ if for all integers j

$$|f^j(x) - y_j| \leq \epsilon .$$

DEFINITION 7.9. A compact invariant set S is hyperbolic if there are positive constants K, α and a projection matrix valued function $P(x)$, $x \in S$, of constant rank such that for all x in S

$$P(f(x))Df(x) = Df(x)P(x) ,$$

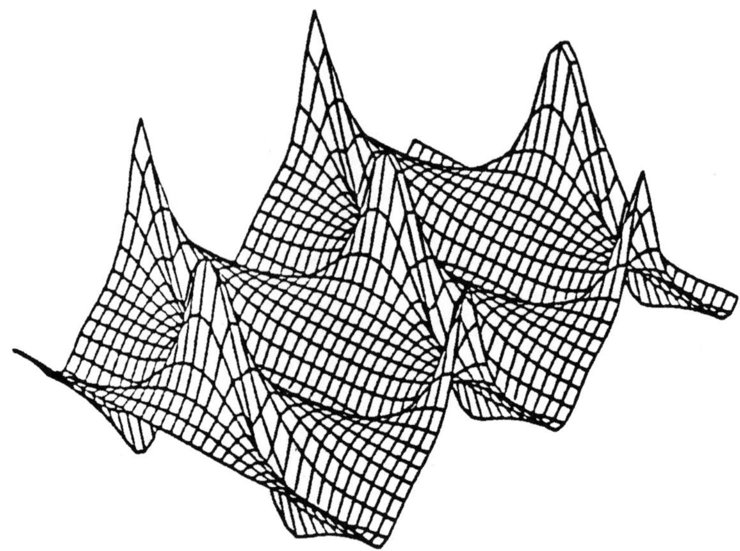

FIGURE 5. Spatial-temporal profile realization of L_{out} in Figure 2 (coordinates are the same as in Figure 1).

$$|Df^j(x)P(x)| \le Ke^{-\alpha j} , \quad (j \ge 0) ,$$

$$|Df^j(x)(I - P(x))| \le Ke^{\alpha j} , \quad (j \le 0) .$$

THEOREM 7.10 (Shadowing Lemma [**173**]). *Let S be a compact hyperbolic set for the C^1 diffeomorphism $f : \mathbb{R}^n \mapsto \mathbb{R}^n$. Then given $\epsilon > 0$ sufficiently small there exists $\delta > 0$ such that every δ pseudo-orbit in S has a unique ϵ-shadowing orbit.*

The proof of this theorem by Palmer [**173**] is overall a fixed point argument with the help of Green functions for linear maps.

Let y_0 be a transversal homoclinic point asymptotic to a saddle x_0 of a C^1 diffeomorphism $f : \mathbb{R}^n \mapsto \mathbb{R}^n$. Then the set

$$S = \{x_0\} \cup \{f^j(y_0) : j \in Z\}$$

is hyperbolic. Denote by A_0 and A_1 the two orbit segments of length $2m + 1$

$$A_0 = \{x_0, x_0, \cdots , x_0\} , \quad A_1 = \{f^{-m}(y_0), f^{-m+1}(y_0), \cdots , f^{m-1}(y_0), f^m(y_0)\} .$$

Let

$$a = (\cdots , a_{-1}, a_0, a_1, \cdots) ,$$

where $a_j \in \{0, 1\}$, be any doubly infinite binary sequence. Let A be the doubly infinite sequence of points in S, associated with a

$$A = \{\cdots , A_{a_{-1}}, A_{a_0}, A_{a_1}, \cdots\} .$$

When m is sufficiently large, A is a δ pseudo-orbit in S. By the shadowing lemma (Theorem 7.10), there is a unique ϵ-shadowing orbit that shadows A. In this manner, Palmer [**173**] gave a beautiful proof of Smale's horseshoe theorem.

DEFINITION 7.11. Denote by Σ the set of doubly infinite binary sequences

$$a = (\cdots , a_{-1}, a_0, a_1, \cdots) ,$$

where $a_j \in \{0, 1\}$. We give the set $\{0, 1\}$ the discrete topology and Σ the product topology. The Bernoulli shift χ is defined by

$$[\chi(a)]_j = a_{j+1} \ .$$

THEOREM 7.12. *Let y_0 be a transversal homoclinic point asymptotic to a saddle x_0 of a C^1 diffeomorphism $f : \mathbb{R}^n \mapsto \mathbb{R}^n$. Then there is a homeomorphism ϕ of Σ onto a compact subset of $\mathbb{R}^n$ which is invariant under f and such that when m is sufficiently large*

$$f^{2m+1} \circ \phi = \phi \circ \chi \ ,$$

that is, the action of f^{2m+1} on $\phi(\Sigma)$ is topologically conjugate to the action of χ on Σ.

Here one can define $\phi(a)$ to be the point on the shadowing orbit that shadows the midpoint of the orbit segment A_{a_0}, which is either x_0 or y_0. The topological conjugacy can be easily verified. For details, see [**173**]. Other references can be found in [**172**] [**213**].

7.7. Shadowing Lemma and Chaos in Infinite-D Periodic Systems

An infinite-dimensional periodic system defined in a Banach space X can be written in the general form

$$\dot{x} = F(x, t) \ ,$$

where $x \in X$, and $F(x, t)$ is periodic in t. Let f be the Poincaré period map. When f is a C^1 map which needs not to be invertible, shadowing lemma and symbolic dynamics around a transversal homoclinic orbit can both be established [**193**] [**194**] [**83**]. Other references can be found in [**79**] [**43**] [**214**] [**30**]. There exists also a work on horseshoe construction without shadowing lemma for sinusoidally forced vibrations of buckled beam [**86**].

7.8. Periodically Perturbed Sine-Gordon (SG) Equation

We continue from section 6.9, and use the notations in section 7.6. For the periodically perturbed sine-Gordon equation (6.51), the Poincaré period map is a C^1 diffeomorphism in H^1. As a corollary of the result in last section, we have the theorem on the existence of chaos.

THEOREM 7.13. *There is an integer m and a homeomorphism ϕ of Σ onto a compact Cantor subset Λ of H^1. Λ is invariant under the Poincaré period-2π map P of the periodically perturbed sine-Gordon equation (6.51). The action of P^{2m+1} on Λ is topologically conjugate to the action of χ on Σ : $P^{2m+1} \circ \phi = \phi \circ \chi$. That is, the following diagram commutes:*

$$
\begin{array}{ccc}
\Sigma & \xrightarrow{\ \phi\ } & \Lambda \\
\chi \downarrow & & \downarrow P^{2m+1} \\
\Sigma & \xrightarrow[\ \phi\ ]{} & \Lambda
\end{array}
$$

7.9. Shadowing Lemma and Chaos in Finite-D Autonomous Systems

A finite-dimensional autonomous system can be written in the general form

$$\dot{x} = F(x) \ ,$$

where $x \in \mathbb{R}^n$. In this case, a transversal homoclinic orbit can be an orbit asymptotic to a normally hyperbolic limit cycle. That is, it is an orbit in the intersection

of the stable and unstable manifolds of a normally hyperbolic limit cycle. Instead
of the Poincaré period map as for periodic system, one may want to introduce the
so-called Poincaré return map which is a map induced by the flow on a codimension 1 section which is transversal to the limit cycle. Unfortunately, such a map
is not even well-defined in the neighborhood of the homoclinic orbit. This poses a
challenging difficulty in extending the arguments as in the case of a Poincaré period
map. In 1996, Palmer [174] completed a proof of a shadowing lemma and existence
of chaos using Newton's method. It will be difficult to extend this method to infinite dimensions, since it used heavily differentiations in time. Other references can
be found in [50] [51] [186] [71].

7.10. Shadowing Lemma and Chaos in Infinite-D Autonomous Systems

An infinite-dimensional autonomous system defined in a Banach space X can
be written in the general form

$$\dot{x} = F(x) \ ,$$

where $x \in X$. In 2002, the author [125] completed a proof of a shadowing lemma
and existence of chaos using Fenichel's persistence of normally hyperbolic invariant
manifold idea. The setup is as follows,

- **Assumption (A1):** There exist a hyperbolic limit cycle S and a transversal homoclinic orbit ξ asymptotic to S. As curves, S and ξ are C^3.
- **Assumption (A2):** The Fenichel fiber theorem is valid at S. That is,
  there exist a family of unstable Fenichel fibers $\{\mathcal{F}^u(q) : q \in S\}$ and a
  family of stable Fenichel fibers $\{\mathcal{F}^s(q) : q \in S\}$. For each fixed $q \in S$,
  $\mathcal{F}^u(q)$ and $\mathcal{F}^s(q)$ are C^3 submanifolds. $\mathcal{F}^u(q)$ and $\mathcal{F}^s(q)$ are C^2 in $q, \forall q \in$
  S. The unions $\bigcup_{q\in S} \mathcal{F}^u(q)$ and $\bigcup_{q\in S} \mathcal{F}^s(q)$ are the unstable and stable
  manifolds of S. Both families are invariant, i.e.

  $$F^t(\mathcal{F}^u(q)) \subset \mathcal{F}^u(F^t(q)), \forall\, t \leq 0, q \in S,$$

  $$F^t(\mathcal{F}^s(q)) \subset \mathcal{F}^s(F^t(q)), \forall\, t \geq 0, q \in S,$$

  where F^t is the evolution operator. There are positive constants κ and $\widehat{C}$
  such that $\forall q \in S$, $\forall q^- \in \mathcal{F}^u(q)$ and $\forall q^+ \in \mathcal{F}^s(q)$,

  $$\|F^t(q^-) - F^t(q)\| \leq \widehat{C}e^{\kappa t}\|q^- - q\|, \forall\, t \leq 0 \ ,$$

  $$\|F^t(q^+) - F^t(q)\| \leq \widehat{C}e^{-\kappa t}\|q^+ - q\|, \forall\, t \geq 0 \ .$$

- **Assumption (A3):** $F^t(q)$ is C^0 in t, for $t \in (-\infty, \infty)$, $q \in X$. For any
  fixed $t \in (-\infty, \infty)$, $F^t(q)$ is a C^2 diffeomorphism on X.

REMARK 7.14. Notice that we do not assume that as functions of time, S and
ξ are C^3 , and we only assume that as curves, S and ξ are C^3.

Under the above setup, a shadowing lemma and existence of chaos can be
proved [125]. Another crucial element in the argument is the establishment of a
λ-lemma (also called inclination lemma) which will be discussed in a later section.

7.11. A Derivative Nonlinear Schrödinger Equation

We continue from section 6.10, and consider the derivative nonlinear Schrödinger equation (6.52). The transversal homoclinic orbit given in Theorem 6.8 is a classical solution. Thus, Assumption (A1) is valid. Assumption (A2) follows from the standard arguments in [139] [136] [124]. Since the perturbation in (6.52) is bounded, Assumption (A3) follows from standard arguments. Thus there exists chaos in the derivative nonlinear Schrödinger equation (6.52) [125].

7.12. λ-Lemma

λ-Lemma (inclination lemma) has been utilized in proving many significant theorems in dynamical system [171] [170] [197]. Here is another example, in [125], it is shown that λ-Lemma is crucial for proving a shadowing lemma in an infinite dimensional autonomous system. Below we give a brief introduction to λ-Lemma [171].

Let f be a C^r $(r \geq 1)$ diffeomorphism in $\mathbb{R}^m$ with 0 as a saddle. Let E^s and E^u be the stable and unstable subspaces. Through changing coordinates, one can obtain that for the local stable and unstable manifolds,

$$W_{\mathrm{loc}}^s \subset E^s , \quad W_{\mathrm{loc}}^u \subset E^u .$$

Let $B^s \subset W_{\mathrm{loc}}^s$ and $B^u \subset W_{\mathrm{loc}}^u$ be balls, and $V = B^s \times B^u$. Let $q \in W_{\mathrm{loc}}^s/\{0\}$, and D^u be a disc of the same dimension as B^u, which is transversal to W_{loc}^s at q.

LEMMA 7.15 (The λ-Lemma). *Let D_n^u be the connected component of $f^n(D^u) \cap V$ to which $f^n(q)$ belongs. Given $\epsilon > 0$, there exists n_0 such that if $n > n_0$, then D_n^u is ϵ C^1-close to B^u.*

The proof of the lemma is not complicated, yet nontrivial [171]. In different situations, the claims of the λ-lemma needed may be different [171] [125]. Nevertheless, the above simple version illustrated the spirit of λ-lemmas.

7.13. Homoclinic Tubes and Chaos Cascades

When studying high-dimensional systems, instead of homoclinic orbits one is more interested in the so-called homoclinic tubes [126] [127] [119] [128] [130]. The concept of a homoclinic tube was introduced by Silnikov [187] in a study on the structure of the neighborhood of a homoclinic tube asymptotic to an invariant torus σ under a diffeomorphism F in a finite dimensional phase space. The asymptotic torus is of saddle type. The homoclinic tube consists of a doubly infinite sequence of tori $\{\sigma_j, \ j = 0, \pm 1, \pm 2, \cdots\}$ in the transversal intersection of the stable and unstable manifolds of σ, such that $\sigma_{j+1} = F \circ \sigma_j$ for any j. It is a generalization of the concept of a transversal homoclinic orbit when the points are replaced by tori.

We are interested in homoclinic tubes for several reasons [126] [127] [119] [128] [130]: 1. Especially in high dimensions, dynamics inside each invariant tubes in the neighborhoods of homoclinic tubes are often chaotic too. We call such chaotic dynamics *"chaos in the small"*, and the symbolic dynamics of the invariant tubes *"chaos in the large"*. Such cascade structures are more important than the structures in a neighborhood of a homoclinic orbit, when high or infinite dimensional dynamical systems are studied. 2. Symbolic dynamics structures in the neighborhoods of homoclinic tubes are more observable than in the neighborhoods of homoclinic orbits in numerical and physical experiments. 3. When studying high or infinite dimensional Hamiltonian system (for example, the cubic nonlinear Schrödinger equation under Hamiltonian perturbations), each invariant tube contains both KAM tori and stochastic layers (chaos in the small). Thus, not only dynamics inside each stochastic layer is chaotic, all these stochastic layers also move chaotically under Poincaré maps.

Using the shadowing lemma technique developed in [125], we obtained in [126] a theorem on the symbolic dynamics of submanifolds in a neighborhood of a homoclinic tube under a C^3-diffeomorphism defined in a Banach space. Such a proof removed an uncheckable assumption in [187]. The result of [187] gives a symbolic labeling of all the invariant tubes around a homoclinic tube. Such a symbolic labeling does not imply the symbolic dynamics of a single map proved in [126].

Then in [127], as an example, the following sine-Gordon equation under chaotic perturbation is studied,

$$(7.11) \qquad u_{tt} = c^2 u_{xx} + \sin u + \epsilon[-au + f(t, \theta_2, \theta_3, \theta_4)(\sin u - u)] \,,$$

which is subject to periodic boundary condition and odd constraint

$$(7.12) \qquad u(t, x + 2\pi) = u(t, x) \,, \quad u(t, -x) = -u(t, x) \,,$$

where u is a real-valued function of two real variables $t \geq 0$ and x, c is a parameter, $\frac{1}{2} < c < 1$, $a > 0$ is a parameter, $\epsilon \geq 0$ is a small parameter, f is periodic in $t, \theta_2, \theta_3, \theta_4$, and $(\theta_2, \theta_3, \theta_4) \in \mathbb{T}^3$, by introducing the extra variable $\theta_1 = \omega_1 t + \theta_1^0$, f takes the form

$$f(t) = \sum_{n=1}^{4} a_n \cos[\theta_n(t)] \,,$$

and a_n's are parameters. Let $\theta_n = \omega_n t + \theta_n^0 + \epsilon^\mu \vartheta_n$, $n = 2, 3, 4$, $\mu > 1$, and ϑ_n's are given by the ABC flow [57] which is verified numerically to be chaotic,

$$\begin{aligned}
\dot{\vartheta}_2 &= A \sin \vartheta_4 + C \cos \vartheta_3 \,, \\
\dot{\vartheta}_3 &= B \sin \vartheta_2 + A \cos \vartheta_4 \,, \\
\dot{\vartheta}_4 &= C \sin \vartheta_3 + B \cos \vartheta_2 \,,
\end{aligned}$$

where A, B, and C are real parameters. Existence of a homoclinic tube is proved. As a corollary of the theorem proved in [126], Symbolic dynamics of tori around the homoclinic tube is established, which is the "chaos in the large". The chaotic dynamics of $(\theta_2, \theta_3, \theta_4)$ is the "chaos in the small".

Here we see the embedding of smaller scale chaos in larger scale chaos. By introducing more variables, such embedding can be continued with even smaller scale chaos. This leads to a chain of embeddings. We call this chain of embeddings of smaller scale chaos in larger scale chaos, a "chaos cascade". We hope that such "chaos cascade" will be proved important.

CHAPTER 8

Stabilities of Soliton Equations in $\mathbb{R}^n$

Unlike soliton equations under periodic boundary conditions, phase space structures, especially hyperbolic structures, of soliton equations under decay boundary conditions are not well understood. The application of Bäcklund-Darboux transformations for generating hyperbolic foliations is not known in decay boundary condition case. Nevertheless, we believe that Bäcklund-Darboux transformations may still have great potentials for understanding phase space structures. This can be a great area for interested readers to work in. The common use of Bäcklund-Darboux transformations is to generate multi-soliton solutions from single soliton solutions. The relations between different soliton solutions in phase spaces (for example certain Sobolev spaces) are not clear yet. There are some studies on the linear and nonlinear stabilities of traveling-wave (soliton) solutions [175] [148]. Results concerning soliton equations are only stability results. Instability results are obtained for non-soliton equations. On the other hand, non-soliton equations are not integrable, have no Lax pair structures, and their phase space structures are much more difficult to be understood. The challenging problem here is to identify which soliton equations possess traveling-wave solutions which are unstable. The reason why we emphasize instabilities is that they are the sources of chaos. Below we are going to review two types of studies on the phase space structures of soliton equations under decay boundary conditions.

8.1. Traveling Wave Reduction

If we only consider traveling-wave solutions to the soliton equations or generalized (or perturbed) soliton equations, the resulting equations are ordinary differential equations with parameters. These ODEs often offer physically meaningful and mathematically pleasant problems to be studied. One of the interesting features is that the traveling-wave solutions correspond to homoclinic orbits to such ODEs. We also naturally expect that such ODEs can have chaos. Consider the KdV equation

$$(8.1) \qquad u_t + 6uu_x + u_{xxx} = 0 \, ,$$

and let $u(x,t) = U(\xi)$, $\xi = x - ct$ (c is a real parameter), we have

$$U''' + 6UU' - cU' = 0 \, .$$

After one integration, we get

$$(8.2) \qquad U'' + 3U^2 - cU + c_1 = 0 \, ,$$

where c_1 is a real integration constant. This equation can be rewritten as a system,

$$(8.3) \qquad \begin{cases} U' = V \, , \\ V' = -3U^2 + cU - c_1 \, . \end{cases}$$

85

For example, when $c_1 = 0$ and $c > 0$ we have the soliton,

$$(8.4) \qquad U_s = \frac{c}{2}\,\mathrm{sech}^2[\pm\frac{\sqrt{c}}{2}\xi] \ ,$$

which is a homoclinic orbit asymptotic to 0.

We may have an equation of the following form describing a real physical phenomenon,

$$u_t + [f(u)]_x + u_{xxx} + \epsilon g(u) = 0 \ ,$$

where ϵ is a small parameter, and traveling waves are physically important. Then we have

$$U''' + [f(U)]' - cU' + \epsilon g(U) = 0 \ ,$$

which can be rewritten as a system,

$$\begin{cases} U' = V \ , \\ V' = W \ , \\ W' = -f'V + cV - \epsilon g(U) \ . \end{cases}$$

This system may even have chaos. Such low dimensional systems are very popular in current dynamical system studies and are mathematically very pleasant [94] [95]. Nevertheless, they represent a very narrow class of solutions to the original PDE, and in no way they can describe the entire phase space structures of the original PDE.

8.2. Stabilities of the Traveling-Wave Solutions

The first step toward understanding the phase space structures of soliton equations under decay boundary conditions is to study the linear or nonlinear stabilities of traveling-wave solutions [175] [148] [75] [98]. Consider the KdV equation (8.1), and study the linear stability of the traveling wave (8.4). First we change the variables from (x, t) into (ξ, t) where $\xi = x - ct$; then in terms of the new variables (ξ, t), the traveling wave (8.4) is a fixed point of the KdV equation (8.1). Let $u(x, t) = U_s(\xi) + v(\xi, t)$ and linearize the KdV equation (8.1) around $U_s(\xi)$, we have

$$\partial_t v - c\partial_\xi v + 6[U_s\partial_\xi v + v\partial_\xi U_s] + \partial_\xi^3 v = 0 \ .$$

We seek solutions to this equation in the form $v = e^{\lambda t}Y(\xi)$ and we have the eigenvalue problem:

$$(8.5) \qquad \lambda Y - c\partial_\xi Y + 6[U_s\partial_\xi Y + Y\partial_\xi U_s] + \partial_\xi^3 Y = 0 \ .$$

One method for studying such eigenvalue problem is carried in two steps [175]: First, one studies the $|\xi| \to \infty$ limit system

$$\lambda Y - c\partial_\xi Y + \partial_\xi^3 Y = 0 \ ,$$

and finds that it has solutions $Y(\xi) = e^{\mu_j\xi}$ for $j = 1, 2, 3$ where the μ_j satisfy

$$\mathrm{Re}\{\mu_1\} < 0 < \mathrm{Re}\{\mu_l\} \quad \text{for } l = 2, 3.$$

Thus the equation (8.5) has a one-dimensional subspace of solutions which decay as $x \to \infty$, and a two-dimensional subspace of solutions which decay as $x \to -\infty$. λ will be an eigenvalue when these subspaces meet nontrivially. In step 2, one measures the angle between these subspaces by a Wronskian-like analytic function

$D(\lambda)$, named Evans's function. One interpretation of this function is that it is like a transmission coefficient, in the sense that for the solution of (8.5) satisfying

$$Y(\xi) \sim e^{\mu_1 \xi} \quad \text{as } \xi \to \infty ,$$

we have

$$Y(\xi) \sim D(\lambda) e^{\mu_1 \xi} \quad \text{as } \xi \to -\infty .$$

In equation (8.5), for $\operatorname{Re}\{\lambda\} > 0$, if $D(\lambda)$ vanishes, then λ is an eigenvalue, and conversely. The conclusion for the equation (8.5) is that there is no eigenvalue with $\operatorname{Re}\{\lambda\} > 0$. In fact, for the KdV equation (8.1), U_s is H^1-orbitally stable [26] [116] [199] [117].

For generalized soliton equations, linear instability results have been established [175]. Let $f(u) = u^{p+1}/(p+1)$, and

$$U_s(\xi) = a \operatorname{sech}^{2/p}(b\xi)$$

for appropriate constants a and b. For the generalized KdV equation [175]

$$\partial_t u + \partial_x f(u) + \partial_x^3 u = 0 ,$$

if $p > 4$, then U_s is linearly unstable for all $c > 0$. If $1 < p < 4$, then U_s is H^1-orbitally stable. For the generalized Benjamin-Bona-Mahoney equation [175]

$$\partial_t u + \partial_x u + \partial_x f(u) - \partial_t \partial_x^2 u = 0 ,$$

if $p > 4$, there exists a positive number $c_0(p)$, such that U_s with $1 < c < c_0(p)$ are linearly unstable. For the generalized regularized Boussinesq equation [175]

$$\partial_t^2 u - \partial_x^2 u - \partial_x^2 f(u) - \partial_t^2 \partial_x^2 u = 0 ,$$

if $p > 4$, U_s is linearly unstable if $1 < c^2 < c_0^2(p)$, where

$$c_0^2(p) = 3p/(4 + 2p) .$$

8.3. Breathers

Breathers are periodic solutions of soliton equations under decay boundary conditions. The importance of breathers with respect to the phase space structures of soliton equations is not clear yet. Bäcklund-Darboux transformations may play a role for this. For example, starting from a breather with one frequency, Bäcklund-Darboux transformations can generate a new breather with two frequencies, i.e. the new breather is a quasiperiodic solution. The stability of breathers will be a very important subject to study.

First we study the reciprocal relation between homoclinic solutions generated through Bäcklund-Darboux transformations for soliton equations under periodic boundary conditions and breathers for soliton equations under decay boundary conditions. Consider the sine-Gordon equation

$$(8.6) \qquad\qquad u_{tt} - u_{xx} + \sin u = 0 ,$$

under the periodic boundary condition

$$(8.7) \qquad\qquad u(x + L, t) = u(x, t) ,$$

and the decay boundary condition

$$(8.8) \qquad\qquad u(x, t) \to 0 \quad \text{as } |x| \to \infty .$$

Starting from the trivial solution $u = \pi$, one can generate the homoclinic solution to the Cauchy problem (8.6) and (8.7) through Bäcklund-Darboux transformations [59],

$$(8.9) \qquad u_H(x,t) = \pi + 4\tan^{-1}\left[\frac{\tan\nu\cos[(\cos\nu)x]}{\cosh[(\sin\nu)t]}\right],$$

where $L = 2\pi/\cos\nu$. Applying the sine-Gordon symmetry

$$(x,t,u) \longrightarrow (t,x,u-\pi),$$

one can immediately generate the breather solution to the Cauchy problem (8.6) and (8.8) [59],

$$(8.10) \qquad u_B(x,t) = 4\tan^{-1}\left[\frac{\tan\nu\cos[(\cos\nu)t]}{\cosh[(\sin\nu)x]}\right].$$

The period of this breather is $L = 2\pi/\cos\nu$. The linear stability can be investigated through studying the Floquet theory for the linear partial differential equation which is the linearized equation of (8.6) at u_B (8.10):

$$(8.11) \qquad u_{tt} - u_{xx} + (\cos u_B)u = 0,$$

under the decay boundary condition

$$(8.12) \qquad u(x,t) \to 0 \quad \text{as } |x| \to \infty.$$

Such studies will be very important in terms of understanding the phase space structures in the neighborhood of the breather and developing infinite dimensional Floquet theory. In fact, McLaughlin and Scott [156] had studied the equation (8.11) through infinitessimal Bäcklund transformations, which lead to the so-called radiation of breather phenomenon. In physical variables only, the Bäcklund transformation for the sine-Gordon equation (8.6) is given as,

$$\frac{\partial}{\partial\xi}\left[\frac{u_+ + u_-}{2}\right] = -\frac{i}{\zeta}\sin\left[\frac{u_+ - u_-}{2}\right],$$

$$\frac{\partial}{\partial\eta}\left[\frac{u_+ - u_-}{2}\right] = i\zeta\sin\left[\frac{u_+ + u_-}{2}\right],$$

where $\xi = \frac{1}{2}(x+t)$, $\eta = \frac{1}{2}(x-t)$, and ζ is a complex Bäcklund parameter. Through the nonlinear superposition principle [179] [8],

$$(8.13) \qquad \tan\left[\frac{u_+ - u_-}{4}\right] = \frac{\zeta_1 + \zeta_2}{\zeta_1 - \zeta_2}\tan\left[\frac{u_1 - u_2}{4}\right]$$

built upon the Bianchi diagram [179] [8]. The breather u_B (8.10) is produced by starting with $u_- = 0$, and choosing $\zeta_1 = \cos\nu + i\sin\nu$ and $\zeta_2 = -\cos\nu + i\sin\nu$. Let $\tilde{u}_-$ be a solution to the linearized sine-Gordon equation (8.11) with u_B replaced by $u_- = 0$,

$$\tilde{u}_- = e^{i(kt+\sqrt{k^2-1}x)}.$$

Then variation of the nonlinear superposition principle leads to the solution to the linearized sine-Gordon equation (8.11):

$$\tilde{u}_+ = \tilde{u}_- + \frac{i\alpha}{\beta}\cos^2(u_B/4)\left[\cos^2[(u_1-u_2)/4]\right]^{-1}(\tilde{u}_1 - \tilde{u}_2)$$

$$= (1+A^2)^{-1}\left[a_1 + a_2\frac{\sin^2\beta t}{\cosh^2\alpha x} + \frac{a_3}{2}\frac{(\cosh^2\alpha x)'}{\cosh^2\alpha x} + \frac{a_4}{2}\frac{(\sin^2\beta t)'}{\cosh^2\alpha x}\right]\tilde{u}_-,$$

where

$$\alpha = \sin \nu\,, \quad \beta = \cos \nu\,, \quad A = \frac{\alpha}{\beta}\frac{\sin \beta t}{\cosh \alpha x}\,,$$

$$a_1 = \frac{1}{2}(k^2 - \beta^2) - \alpha^2\,, \quad a_2 = \frac{1}{2}(k^2 - \beta^2)\frac{\alpha^2}{\beta^2} + \alpha^2\,,$$

$$a_3 = i\sqrt{k^2 - 1}\,, \quad a_4 = ik\frac{\alpha^2}{\beta^2}\,,$$

which represents the radiation of the breather.

Next we briefly survey some interesting studies on breathers. The rigidity of sine-Gordon breathers was studied by Birnir, Mckean and Weinstein [29] [28]. Under the scaling

$$x \to x' = (1 + \epsilon a)^{-1/2}x\,, \quad t \to t' = (1 + \epsilon a)^{-1/2}t\,, \quad u \to u' = (1 + \epsilon b)u$$

the sine-Gordon equation (8.6) is transformed into

$$u_{tt} - u_{xx} + \sin u + \epsilon(a \sin u + bu \cos u) + O(\epsilon^2) = 0\,.$$

The breather (8.10) is transformed into a breather for the new equation. The remarkable fact is the rigidity of the breather (8.10) as stated in the following theorem.

THEOREM 8.1 (Birnir, Mckean and Weinstein [29]). *If $f(u)$ vanishes at $u = 0$ and is holomorphic in the open strip $|Re\{u\}| < \pi$, then $u_{tt} - u_{xx} + \sin u + \epsilon f(u) + O(\epsilon^2) = 0$ has breathers $u = u_B + \epsilon \hat{u}_1 + O(\epsilon^2)$, one to each $\sin \nu$ from an open subinterval of $(0, 1/\sqrt{2}]$ issuing smoothly from $u = u_B$ at $\epsilon = 0$, with $\hat{u}_1$ vanishing at $x = \pm\infty$, if and only if $f(u)$ is a linear combination of $\sin u$, $u \cos u$, and/or $1 + 3\cos u - 4\cos(u/2) + 4\cos u \ln \cos(u/4)$.*

In fact, it is proved that sine-Gordon equation (8.6) is the only nonlinear wave equation possessing small analytic breathers [104]. These results lend color to the conjecture that, for the general nonlinear wave equation, breathing is an "arithmetical" phenomenon in the sense that it takes place only for isolated nonlinearities, scaling and other trivialities aside.

CHAPTER 9

Lax Pairs of Euler Equations of Inviscid Fluids

The governing equations for the incompressible viscous fluid flow are the Navier-Stokes equations. Turbulence occurs in the regime of high Reynolds number. By formally setting the Reynolds number equal to infinity, the Navier-Stokes equations reduce to the Euler equations of incompressible inviscid fluid flow. One may view the Navier-Stokes equations with large Reynolds number as a singular perturbation of the Euler equations.

Results of T. Kato show that 2D Navier-Stokes equations are globally well-posed in $C^0([0,\infty); H^s(R^2))$, $s > 2$, and for any $0 < T < \infty$, the mild solutions of the 2D Navier-Stokes equations approach those of the 2D Euler equations in $C^0([0,T]; H^s(R^2))$ [101]. 3D Navier-Stokes equations are locally well-posed in $C^0([0,\tau]; H^s(R^3))$, $s > 5/2$, and the mild solutions of the 3D Navier-Stokes equations approach those of the 3D Euler equations in $C^0([0,\tau]; H^s(R^3))$, where τ depends on the norms of the initial data and the external force [99] [100]. Extensive studies on the inviscid limit have been carried by J. Wu et al. [204] [49] [205] [31]. There is no doubt that mathematical study on Navier-Stokes (Euler) equations is one of the most important mathematical problems. In fact, Clay Mathematics Institute has posted the global well-posedness of 3D Navier-Stokes equations as one of the one million dollars problems.

V. Arnold [15] realized that 2D Euler equations are a Hamiltonian system. Extensive studies on the symplectic structures of 2D Euler equations have been carried by J. Marsden, T. Ratiu et al. [151].

Recently, the author found Lax pair structures for Euler equations [123] [141] [131] [128]. Understanding the structures of solutions to Euler equations is of fundamental interest. Of particular interest is the question on the global well-posedness of 3D Navier-Stokes and Euler equations. Our number one hope is that the Lax pair structures can be useful in investigating the global well-posedness. Our secondary hope is that the Darboux transformation [141] associated with the Lax pair can generate explicit representation of homoclinic structures [121].

The philosophical significance of the existence of Lax pairs for Euler equations is even more important. If one defines integrability of an equation by the existence of a Lax pair, then both 2D and 3D Euler equations are integrable. More importantly, both 2D and 3D Navier-Stokes equations at high Reynolds numbers are singularly perturbed integrable systems. Such a point of view changes our old ideology on Euler and Navier-Stokes equations.

9.1. A Lax Pair for 2D Euler Equation

The 2D Euler equation can be written in the vorticity form,

$$(9.1) \qquad \partial_t \Omega + \{\Psi, \Omega\} = 0 \,,$$

91

where the bracket $\{\ ,\ \}$ is defined as

$$\{f,g\} = (\partial_x f)(\partial_y g) - (\partial_y f)(\partial_x g) \ ,$$

Ω is the vorticity, and Ψ is the stream function given by,

$$u = -\partial_y \Psi \ , \quad v = \partial_x \Psi \ ,$$

and the relation between vorticity Ω and stream function Ψ is,

$$\Omega = \partial_x v - \partial_y u = \Delta \Psi \ .$$

THEOREM 9.1 (Li, [**123**]). *The Lax pair of the 2D Euler equation (9.1) is given as*

$$(9.2) \qquad \left\{ \begin{array}{l} L\varphi = \lambda\varphi \ , \\ \partial_t \varphi + A\varphi = 0 \ , \end{array} \right.$$

where

$$L\varphi = \{\Omega, \varphi\} \ , \quad A\varphi = \{\Psi, \varphi\} \ ,$$

and λ is a complex constant, and φ is a complex-valued function.

9.2. A Darboux Transformation for 2D Euler Equation

Consider the Lax pair (9.2) at $\lambda = 0$, i.e.

$$(9.3) \qquad \{\Omega, p\} = 0 \ ,$$

$$(9.4) \qquad \partial_t p + \{\Psi, p\} = 0 \ ,$$

where we replaced the notation φ by p.

THEOREM 9.2 ([**141**]). *Let $f = f(t,x,y)$ be any fixed solution to the system (9.3, 9.4), we define the Gauge transform G_f:*

$$(9.5) \qquad \tilde{p} = G_f p = \frac{1}{\Omega_x}[p_x - (\partial_x \ln f)p] \ ,$$

and the transforms of the potentials Ω and Ψ:

$$(9.6) \qquad \tilde{\Psi} = \Psi + F \ , \quad \tilde{\Omega} = \Omega + \Delta F \ ,$$

where F is subject to the constraints

$$(9.7) \qquad \{\Omega, \Delta F\} = 0 \ , \quad \{\Delta F, F\} = 0 \ .$$

Then $\tilde{p}$ solves the system (9.3, 9.4) at $(\tilde{\Omega}, \tilde{\Psi})$. Thus (9.5) and (9.6) form the Darboux transformation for the 2D Euler equation (9.1) and its Lax pair (9.3, 9.4).

REMARK 9.3. For KdV equation and many other soliton equations, the Gauge transform is of the form [**152**],

$$\tilde{p} = p_x - (\partial_x \ln f)p \ .$$

In general, Gauge transform does not involve potentials. For 2D Euler equation, a potential factor $\frac{1}{\Omega_x}$ is needed. From (9.3), one has

$$\frac{p_x}{\Omega_x} = \frac{p_y}{\Omega_y} \ .$$

The Gauge transform (9.5) can be rewritten as

$$\tilde{p} = \frac{p_x}{\Omega_x} - \frac{f_x}{\Omega_x} \frac{p}{f} = \frac{p_y}{\Omega_y} - \frac{f_y}{\Omega_y} \frac{p}{f} \ .$$

The Lax pair (9.3, 9.4) has a symmetry, i.e. it is invariant under the transform $(t, x, y) \rightarrow (-t, y, x)$. The form of the Gauge transform (9.5) resulted from the inclusion of the potential factor $\frac{1}{\Omega_x}$, is consistent with this symmetry.

Our hope is to use the Darboux transformation to generate homoclinic structures for 2D Euler equation [121].

9.3. A Lax Pair for Rossby Wave Equation

The Rossby wave equation is

$$\partial_t \Omega + \{\Psi, \Omega\} + \beta \partial_x \Psi = 0 \ ,$$

where $\Omega = \Omega(t, x, y)$ is the vorticity, $\{\Psi, \Omega\} = \Psi_x \Omega_y - \Psi_y \Omega_x$, and $\Psi = \Delta^{-1}\Omega$ is the stream function. Its Lax pair can be obtained by formally conducting the transformation, $\Omega = \tilde{\Omega} + \beta y$, to the 2D Euler equation [123],

$$\{\Omega, \varphi\} - \beta \partial_x \varphi = \lambda \varphi \ , \quad \partial_t \varphi + \{\Psi, \varphi\} = 0 \ ,$$

where φ is a complex-valued function, and λ is a complex parameter.

9.4. Lax Pairs for 3D Euler Equation

The 3D Euler equation can be written in vorticity form,

$$(9.8) \qquad \partial_t \Omega + (u \cdot \nabla)\Omega - (\Omega \cdot \nabla)u = 0 \ ,$$

where $u = (u_1, u_2, u_3)$ is the velocity, $\Omega = (\Omega_1, \Omega_2, \Omega_3)$ is the vorticity, $\nabla = (\partial_x, \partial_y, \partial_z)$, $\Omega = \nabla \times u$, and $\nabla \cdot u = 0$. u can be represented by Ω for example through Biot-Savart law.

THEOREM 9.4. *The Lax pair of the 3D Euler equation (9.8) is given as*

$$(9.9) \qquad \begin{cases} L\phi = \lambda\phi \ , \\ \partial_t \phi + A\phi = 0 \ , \end{cases}$$

where

$$L\phi = (\Omega \cdot \nabla)\phi \ , \quad A\varphi = (u \cdot \nabla)\phi \ ,$$

λ *is a complex constant, and ϕ is a complex scalar-valued function.*

THEOREM 9.5 ([39]). *Another Lax pair of the 3D Euler equation (9.8) is given as*

$$(9.10) \qquad \begin{cases} L\varphi = \lambda\varphi \ , \\ \partial_t \varphi + A\varphi = 0 \ , \end{cases}$$

where

$$L\varphi = (\Omega \cdot \nabla)\varphi - (\varphi \cdot \nabla)\Omega \ , \quad A\varphi = (u \cdot \nabla)\varphi - (\varphi \cdot \nabla)u \ ,$$

λ *is a complex constant, and $\varphi = (\varphi_1, \varphi_2, \varphi_3)$ is a complex 3-vector valued function.*

Our hope is that the infinitely many conservation laws generated by $\lambda \in C$ can provide a priori estimates for the global well-posedness of 3D Navier-Stokes equations, or better understanding on the global well-posedness. For more informations on the topics, see [141].

CHAPTER 10

Linearized 2D Euler Equation at a Fixed Point

To begin an infinite dimensional dynamical system study, we investigate the linearized 2D Euler equation at a fixed point [**122**]. We consider the 2D Euler equation (9.1) under periodic boundary condition in both x and y directions with period 2π. Expanding Ω into Fourier series,

$$\Omega = \sum_{k \in \mathbb{Z}^2/\{0\}} \omega_k \, e^{ik \cdot X} \, ,$$

where $\omega_{-k} = \overline{\omega_k}$, $k = (k_1, k_2)^T$, and $X = (x, y)^T$. The 2D Euler equation can be rewritten as

$$(10.1) \qquad \dot{\omega}_k = \sum_{k=p+q} A(p, q) \, \omega_p \omega_q \, ,$$

where $A(p, q)$ is given by,

$$
\begin{aligned}
(10.2) \qquad A(p, q) &= \frac{1}{2}[|q|^{-2} - |p|^{-2}](p_1 q_2 - p_2 q_1) \\[2mm]
&= \frac{1}{2}[|q|^{-2} - |p|^{-2}] \begin{vmatrix} p_1 & q_1 \\ p_2 & q_2 \end{vmatrix} \, ,
\end{aligned}
$$

where $|q|^2 = q_1^2 + q_2^2$ for $q = (q_1, q_2)^T$, similarly for p. The 2D Euler equation (10.1) has huge dimensional equilibrium manifolds.

PROPOSITION 10.1. *For any $k \in Z^2/\{0\}$, the infinite dimensional space*

$$E_k^1 \equiv \left\{ \{\omega_{k'}\} \ \middle| \ \omega_{k'} = 0, \ if \ k' \neq rk, \ \forall r \in R \right\} \, ,$$

and the finite dimensional space

$$E_k^2 \equiv \left\{ \{\omega_{k'}\} \ \middle| \ \omega_{k'} = 0, \ if \ |k'| \neq |k| \right\} \, ,$$

entirely consist of fixed points of the system (10.1).

Fig.1 shows an example on the locations of the modes ($k' = rk$) and ($|k'| = |k|$) in the definitions of E_k^1 and E_k^2 (Proposition 10.1).

10.1. Hamiltonian Structure of 2D Euler Equation

For any two functionals F_1 and F_2 of $\{\omega_k\}$, define their Lie-Poisson bracket:

$$(10.3) \qquad \{F_1, F_2\} = \sum_{k+p+q=0} \begin{vmatrix} q_1 & p_1 \\ q_2 & p_2 \end{vmatrix} \omega_k \, \frac{\partial F_1}{\partial \overline{\omega_p}} \frac{\partial F_2}{\partial \overline{\omega_q}} \, .$$

95

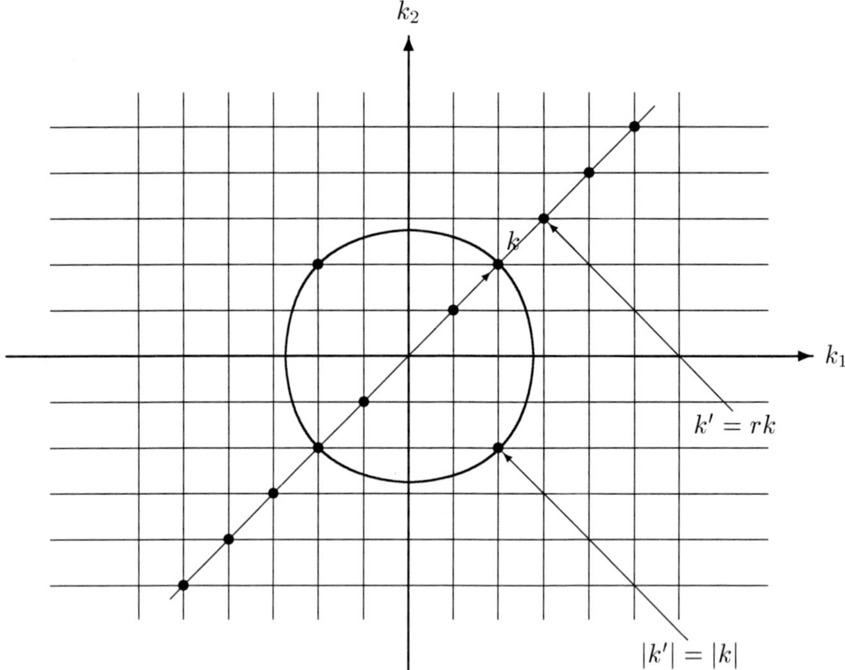

FIGURE 1. An illustration on the locations of the modes ($k' = rk$) and ($|k'| = |k|$) in the definitions of E_k^1 and E_k^2 (Proposition 10.1).

Then the 2D Euler equation (10.1) is a Hamiltonian system [**15**],

$$(10.4) \qquad \dot{\omega}_k = \{\omega_k, H\},$$

where the Hamiltonian H is the kinetic energy,

$$(10.5) \qquad H = \frac{1}{2} \sum_{k \in Z^2/\{0\}} |k|^{-2} |\omega_k|^2.$$

Following are Casimirs (i.e. invariants that Poisson commute with any functional) of the Hamiltonian system (10.4):

$$(10.6) \qquad J_n = \sum_{k_1 + \cdots + k_n = 0} \omega_{k_1} \cdots \omega_{k_n}.$$

10.2. Linearized 2D Euler Equation at a Unimodal Fixed Point

Denote $\{\omega_k\}_{k \in \mathbb{Z}^2/\{0\}}$ by ω. We consider the simple fixed point ω^* [**122**]:

$$(10.7) \qquad \omega_p^* = \Gamma, \quad \omega_k^* = 0, \text{ if } k \neq p \text{ or } -p,$$

of the 2D Euler equation (10.1), where Γ is an arbitrary complex constant. The *linearized two-dimensional Euler equation* at ω^* is given by,

$$(10.8) \qquad \dot{\omega}_k = A(p, k - p) \, \Gamma \, \omega_{k-p} + A(-p, k + p) \, \bar{\Gamma} \, \omega_{k+p} \,.$$

DEFINITION 10.2 (Classes). For any $\hat{k} \in \mathbb{Z}^2/\{0\}$, we define the class $\Sigma_{\hat{k}}$ to be the subset of $\mathbb{Z}^2/\{0\}$:

$$\Sigma_{\hat{k}} = \left\{ \hat{k} + np \in \mathbb{Z}^2/\{0\} \,\middle|\, n \in Z, \ p \text{ is specified in } (10.7) \right\}.$$

See Fig.2 for an illustration of the classes. According to the classification defined in Definition 10.2, the linearized two-dimensional Euler equation (10.8) decouples into infinitely many *invariant subsystems*:

$$\dot{\omega}_{\hat{k}+np} = A(p, \hat{k} + (n-1)p) \, \Gamma \, \omega_{\hat{k}+(n-1)p}$$

(10.9)
$$+ A(-p, \hat{k} + (n+1)p) \, \bar{\Gamma} \, \omega_{\hat{k}+(n+1)p} \, .$$

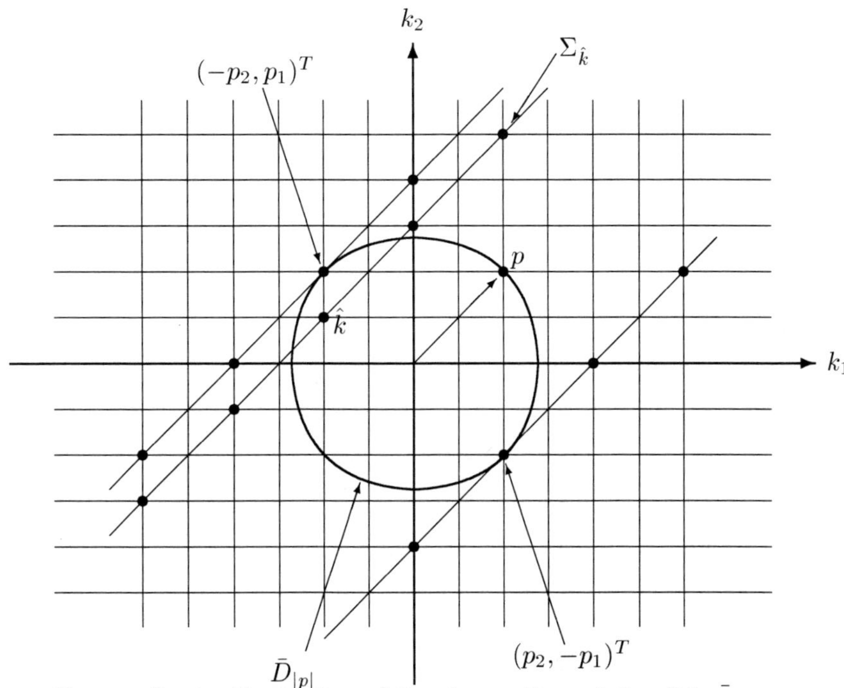

FIGURE 2. An illustration of the classes $\Sigma_{\hat{k}}$ and the disk $\bar{D}_{|p|}$.

10.2.1. Linear Hamiltonian Systems. Each invariant subsystem (10.9) can be rewritten as a linear Hamiltonian system as shown below.

DEFINITION 10.3 (The Quadratic Hamiltonian). The quadratic Hamiltonian $\mathcal{H}_{\hat{k}}$ is defined as:

$$\mathcal{H}_{\hat{k}} = -2 \, \text{Im} \left\{ \sum_{n \in Z} \rho_n \, \Gamma \, A(p, \hat{k} + (n-1)p) \, \omega_{\hat{k}+(n-1)p} \, \bar{\omega}_{\hat{k}+np} \right\}$$

(10.10)
$$= - \begin{vmatrix} p_1 & \hat{k}_1 \\ p_2 & \hat{k}_2 \end{vmatrix} \text{Im} \left\{ \sum_{n \in Z} \Gamma \, \rho_n \, \rho_{n-1} \, \omega_{\hat{k}+(n-1)p} \, \bar{\omega}_{\hat{k}+np} \right\},$$

where $\rho_n = [|\hat{k} + np|^{-2} - |p|^{-2}]$, " Im " denotes " imaginary part ".

Then the invariant subsystem (10.9) can be rewritten as a linear Hamiltonian system [**122**],

$$(10.11) \qquad i\, \dot{\omega}_{\hat{k}+np} = \rho_n^{-1} \, \frac{\partial \mathcal{H}_{\hat{k}}}{\partial \bar{\omega}_{\hat{k}+np}} \; .$$

For a finite dimensional linear Hamiltonian system, it is well-known that the eigenvalues are of four types: real pairs $(c, -c)$, purely imaginary pairs $(id, -id)$, quadruples $(\pm c \pm id)$, and zero eigenvalues [**177**] [**142**] [**17**]. There is also a complete theorem on the normal forms of such Hamiltonians [**17**]. For the above infinite dimensional system (10.9), the classical proofs can not be applied. Nevertheless, the conclusion is still true with different proof [**122**] [**134**]. Let $\mathcal{L}_{\hat{k}}$ be the linear operator defined by the right hand side of (10.9), and H^s be the Sobolev space where $s \geq 0$ is an integer and $H^0 = \ell_2$.

THEOREM 10.4 ([**122**] [**134**]). *The eigenvalues of the linear operator $\mathcal{L}_{\hat{k}}$ in H^s are of four types: real pairs $(c, -c)$, purely imaginary pairs $(id, -id)$, quadruples $(\pm c \pm id)$, and zero eigenvalues.*

10.2.2. Liapunov Stability.

DEFINITION 10.5 (An Important Functional). For each invariant subsystem (10.9), we define the functional $I_{\hat{k}}$,

$$
(10.12) \qquad
\begin{aligned}
I_{\hat{k}} \;&=\; I_{(\text{restricted to } \Sigma_{\hat{k}})} \\[2mm]
&=\; \sum_{n \in Z} \{ |\hat{k} + np|^{-2} - |p|^{-2} \} \left| \omega_{\hat{k}+np} \right|^2 .
\end{aligned}
$$

LEMMA 10.6. *$I_{\hat{k}}$ is a constant of motion for the system (10.11).*

DEFINITION 10.7 (The Disk). The disk of radius $|p|$ in $\mathbb{Z}^2/\{0\}$, denoted by $\bar{D}_{|p|}$, is defined as

$$\bar{D}_{|p|} = \left\{ k \in \mathbb{Z}^2/\{0\} \;\middle|\; |k| \leq |p| \right\}.$$

See Fig.2 for an illustration.

THEOREM 10.8 (Unstable Disk Theorem, [**122**]). *If $\Sigma_{\hat{k}} \cap \bar{D}_{|p|} = \emptyset$, then the invariant subsystem (10.9) is Liapunov stable for all $t \in R$, in fact,*

$$\sum_{n \in Z} \left| \omega_{\hat{k}+np}(t) \right|^2 \leq \sigma \sum_{n \in Z} \left| \omega_{\hat{k}+np}(0) \right|^2 , \qquad \forall t \in R,$$

where

$$\sigma = \left[\max_{n \in Z} \{ -\rho_n \} \right] \left[\min_{n \in Z} \{ -\rho_n \} \right]^{-1} , \qquad 0 < \sigma < \infty.$$

10.2.3. Spectral Theorems.

Again denote by $\mathcal{L}_{\hat{k}}$ the linear operator defined by the right hand side of (10.9).

THEOREM 10.9 (The Spectral Theorem, [**122**] [**134**]). *We have the following claims on the spectrum of the linear operator $\mathcal{L}_{\hat{k}}$:*

(1) If $\Sigma_{\hat{k}} \cap \bar{D}_{|p|} = \emptyset$, then the entire H^s spectrum of the linear operator $\mathcal{L}_{\hat{k}}$ is its continuous spectrum. See Figure 3, where $b = -\frac{1}{2}|\Gamma||p|^{-2} \begin{vmatrix} p_1 & \hat{k}_1 \\ p_2 & \hat{k}_2 \end{vmatrix}$. That is, both the residual and the point spectra of $\mathcal{L}_{\hat{k}}$ are empty.

(2) If $\Sigma_{\hat{k}} \cap \bar{D}_{|p|} \neq \emptyset$, then the entire essential H^s spectrum of the linear operator $\mathcal{L}_{\hat{k}}$ is its continuous spectrum. That is, the residual spectrum of $\mathcal{L}_{\hat{k}}$ is empty. The point spectrum of $\mathcal{L}_{\hat{k}}$ is symmetric with respect to both real and imaginary axes. See Figure 4.

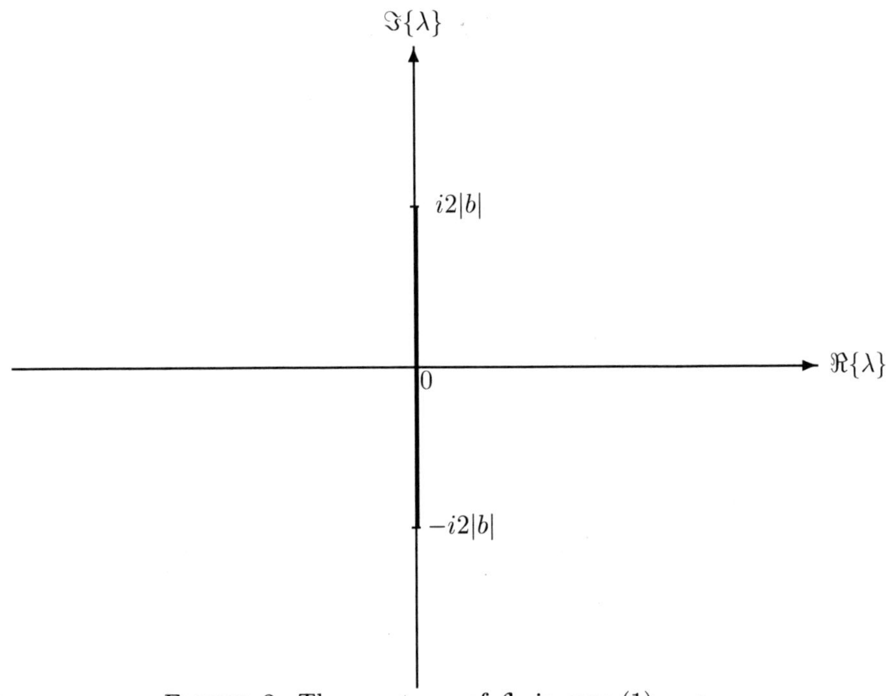

FIGURE 3. The spectrum of $\mathcal{L}_{\hat{k}}$ in case (1).

For a detailed proof of this theorem, see [122] [134]. Denote by L the right hand side of (10.8), i.e. the whole linearized 2D Euler operator, the spectral mapping theorem holds.

THEOREM 10.10 ([112]).

$$\sigma(e^{tL}) = e^{t\sigma(L)}, t \neq 0.$$

10.2.4. A Continued Fraction Calculation of Eigenvalues. Since the introduction of continued fractions for calculating the eigenvalues of steady fluid flow, by Meshalkin and Sinai [161], this topics had been extensively explored [208] [143] [144] [145] [146] [147] [22] [122]. Rigorous justification on the continued fraction calculation was given in [122] [147].

Rewrite the equation (10.9) as follows,

$$(10.13) \qquad \rho_n^{-1}\dot{\tilde{z}}_n = a\left[\tilde{z}_{n+1} - \tilde{z}_{n-1}\right],$$

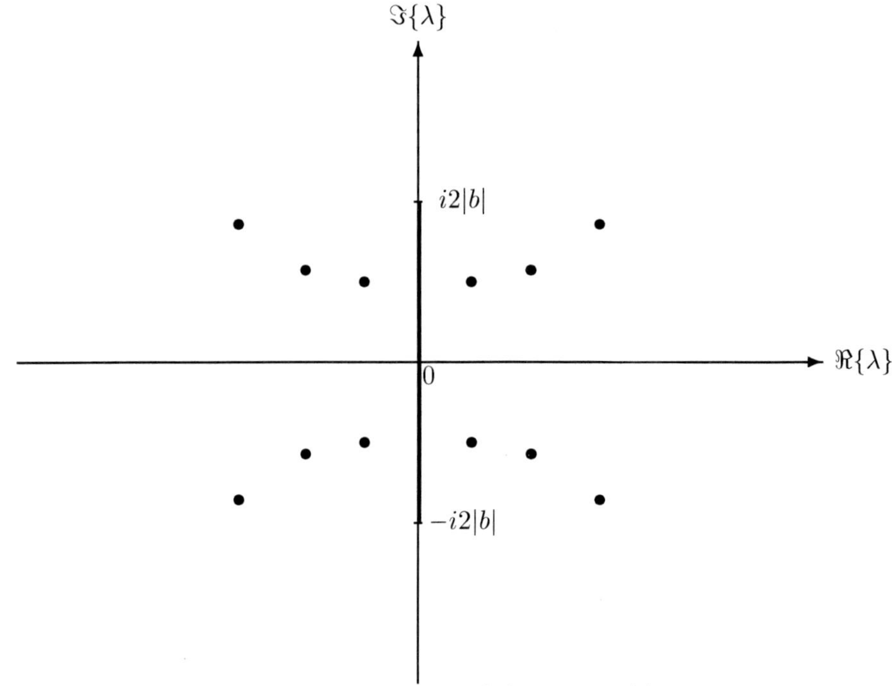

FIGURE 4. The spectrum of $\mathcal{L}_{\hat{k}}$ in case (2).

where $\tilde{z}_n = \rho_n e^{in(\theta+\pi/2)} \omega_{\hat{k}+np}$, $\theta + \gamma = \pi/2$, $\Gamma = |\Gamma| e^{i\gamma}$, $a = \frac{1}{2}|\Gamma| \begin{vmatrix} p_1 & \hat{k}_1 \\ p_2 & \hat{k}_2 \end{vmatrix}$, $\rho_n = |\hat{k} + np|^{-2} - |p|^{-2}$. Let $\tilde{z}_n = e^{\lambda t} z_n$, where $\lambda \in C$; then z_n satisfies

$$(10.14) \qquad a_n z_n + z_{n-1} - z_{n+1} = 0 \, ,$$

where $a_n = \lambda(a\rho_n)^{-1}$. Let $w_n = z_n/z_{n-1}$ [161]; then w_n satisfies

$$(10.15) \qquad a_n + \frac{1}{w_n} = w_{n+1} \, .$$

Iteration of (10.15) leads to the continued fraction solution [161],

$$(10.16) \qquad w_n^{(1)} = a_{n-1} + \cfrac{1}{a_{n-2} + \cfrac{1}{a_{n-3}+\cfrac{1}{\ddots}}} \, .$$

Rewrite (10.15) as follows,

$$(10.17) \qquad w_n = \frac{1}{-a_n + w_{n+1}} \, .$$

Iteration of (10.17) leads to the continued fraction solution [161],

$$(10.18) \qquad w_n^{(2)} = -\cfrac{1}{a_n + \cfrac{1}{a_{n+1}+\cfrac{1}{a_{n+2}+\cfrac{1}{\ddots}}}} \, .$$

The eigenvalues are given by the condition $w_1^{(1)} = w_1^{(2)}$, i.e.

$$(10.19) \qquad f = a_0 + \left(\cfrac{1}{a_{-1} + \cfrac{1}{a_{-2} + \cfrac{1}{a_{-3} + \ddots}}} \right) + \left(\cfrac{1}{a_1 + \cfrac{1}{a_2 + \cfrac{1}{a_3 + \ddots}}} \right) = 0 \, ,$$

where $f = f(\tilde{\lambda}, \hat{k}, p)$, $\tilde{\lambda} = \lambda/a$.

As an example, we take $p = (1,1)^T$. When $\Gamma \neq 0$, the fixed point has 4 eigenvalues which form a quadruple. These four eigenvalues appear in the invariant linear subsystem labeled by $\hat{k} = (-3, -2)^T$. One of them is [122]:

$$(10.20) \qquad \tilde{\lambda} = 2\lambda/|\Gamma| = 0.24822302478255 + i\, 0.35172076526520 \, .$$

See Figure 5 for an illustration. The essential spectrum (= continuous spectrum) of $\mathcal{L}_{\hat{k}}$ with $\hat{k} = (-3, -2)^T$ is the segment on the imaginary axis shown in Figure 5, where $b = -\frac{1}{4}\Gamma$. The essential spectrum (= continuous spectrum) of the linear 2D Euler operator at this fixed point is the entire imaginary axis. Denote by L the

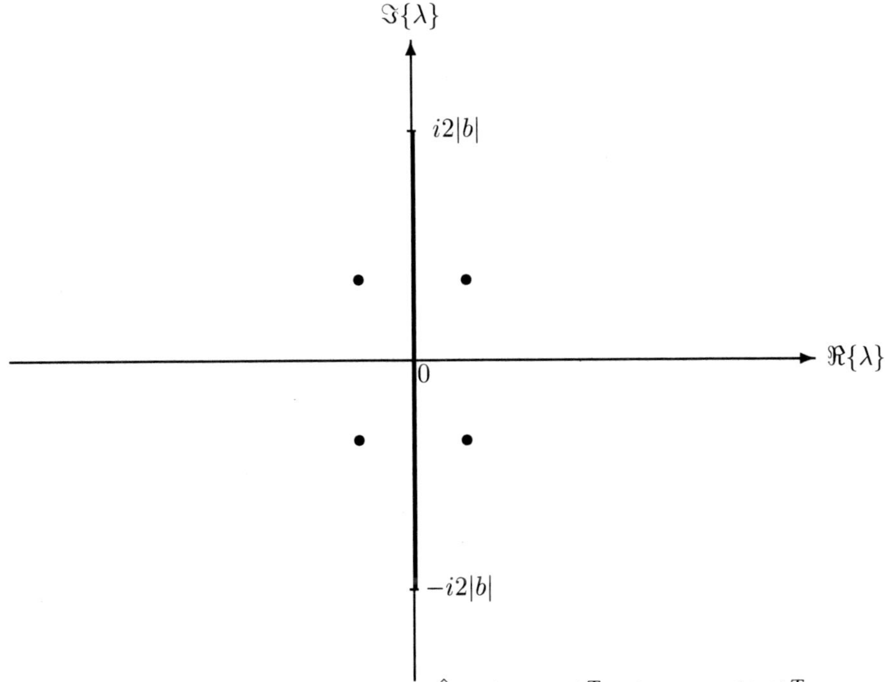

FIGURE 5. The spectrum of $\mathcal{L}_{\hat{k}}$ with $\hat{k} = (-3, -2)^T$, when $p = (1,1)^T$.

right hand side of (10.8), i.e. the whole linearized 2D Euler operator. Let ζ denote the number of points $q \in \mathbb{Z}^2/\{0\}$ that belong to the open disk of radius $|p|$, and such that q is not parallel to p.

THEOREM 10.11 ([112]). *The number of nonimaginary eigenvalues of L (counting the multiplicities) does not exceed 2ζ.*

Another interesting discussion upon the discrete spectrum can be found in [61].

A rather well-known open problem is proving the existence of unstable, stable, and center manifolds. The main difficulty comes from the fact that the nonlinear term is non-Lipschitzian.

CHAPTER 11

Arnold's Liapunov Stability Theory

11.1. A Brief Summary

Let D be a region on the (x, y)-plane bounded by the curves Γ_i $(i = 1, 2)$, an ideal fluid flow in D is governed by the 2D Euler equation written in the stream-function form:

$$(11.1) \qquad \frac{\partial}{\partial t} \Delta \psi = [\nabla \psi, \nabla \Delta \psi] ,$$

where

$$[\nabla \psi, \nabla \Delta \psi] = \frac{\partial \psi}{\partial x} \frac{\partial \Delta \psi}{\partial y} - \frac{\partial \psi}{\partial y} \frac{\partial \Delta \psi}{\partial x} ,$$

with the boundary conditions,

$$\psi|_{\Gamma_i} = c_i(t) , \quad c_1 \equiv 0 , \quad \frac{d}{dt} \oint_{\Gamma_i} \frac{\partial \psi}{\partial n} ds = 0 .$$

For every function $f(z)$, the functional

$$(11.2) \qquad F = \int \int_D f(\Delta \psi) \, dx dy$$

is a constant of motion (a Casimir) for (11.1). The conditional extremum of the kinetic energy

$$(11.3) \qquad E = \frac{1}{2} \int \int_D \nabla \psi \cdot \nabla \psi \, dx dy$$

for fixed F is given by the Lagrange's formula [14],

$$(11.4) \qquad \delta H = \delta(E + \lambda F) = 0 , \quad \Rightarrow \quad \psi_0 = \lambda f'(\Delta \psi_0) .$$

where λ is the Lagrange multiplier. Thus, ψ_0 is the stream function of a stationary flow, which satisfies

$$(11.5) \qquad \psi_0 = \Phi(\Delta \psi_0) ,$$

where $\Phi = \lambda f'$. The second variation is given by [14],

$$(11.6) \qquad \delta^2 H = \frac{1}{2} \int \int_D \left\{ \nabla \phi \cdot \nabla \phi + \Phi'(\Delta \psi_0) \, (\Delta \phi)^2 \right\} dx dy .$$

Let $\psi = \psi_0 + \varphi$ be a solution to the 2D Euler equation (11.1), Arnold proved the estimates [16]: (a). when $c \leq \Phi'(\Delta \psi_0) \leq C, 0 < c \leq C < \infty,$

$$\int \int_D \left\{ \nabla \varphi(t) \cdot \nabla \varphi(t) + c[\Delta \varphi(t)]^2 \right\} dx dy \leq \int \int_D \left\{ \nabla \varphi(0) \cdot \nabla \varphi(0) + C[\Delta \varphi(0)]^2 \right\} dx dy,$$

103

for all $t \in (-\infty, +\infty)$, (b). when $c \le -\Phi'(\Delta\psi_0) \le C$, $0 < c < C < \infty$,

$$\int\int_D \left\{ c[\Delta\varphi(t)]^2 - \nabla\varphi(t)\cdot\nabla\varphi(t) \right\} dxdy \le \int\int_D \left\{ C[\Delta\varphi(0)]^2 - \nabla\varphi(0)\cdot\nabla\varphi(0) \right\} dxdy,$$

for all $t \in (-\infty, +\infty)$. Therefore, when the second variation (11.6) is positive definite, or when

$$\int\int_D \left\{ \nabla\phi \cdot \nabla\phi + [\max \Phi'(\Delta\psi_0)] (\Delta\phi)^2 \right\} dxdy$$

is negative definite, the stationary flow (11.5) is nonlinearly stable (Liapunov stable).

Arnold's Liapunov stability theory had been extensively explored, see e.g. [85] [151] [202].

11.2. Miscellaneous Remarks

Establishing Liapunov instability along the line of the above, has not been successful. Some rather technical, with no clear physical meaning as above, argument showing nonlinear instability starting from linear instability, has been established by Guo et al. [77] [7] [72].

Yudovich [209] had been promoting the importance of the so-called "slow collapse", that is, not finite time blowup, rather growing to infinity in time. The main thought is that if derivatives growing to infinity in time, the function itself should gain randomness. Yudovich [208] had been studying bifurcations of fluid flows.

There are also interests [74] in studying structures of divergent free 2D vector fields on 2-tori.

CHAPTER 12

Miscellaneous Topics

This chapter serves as a guide to other interesting topics. Some topics are already well-developed. Others are poorly developed in terms of partial differential equations.

12.1. KAM Theory

KAM (Kolmogorov-Arnold-Moser) theory in finite dimensions has been a well-known topic [11]. It dealt with the persistence of Liouville tori in integrable Hamiltonian systems under Hamiltonian perturbations. A natural idea of constructing such tori in perturbed systems is conducting canonical transformations which lead to small divisor problem. To overcome such difficulties, Kolmogorov introduced the Newton's method to speed up the rate of convergence of the canonical transformation series. Under certain non-resonance condition and non-degeneracy condition of certain Hessian, Arnold completed the proof of a rather general theorem [11]. Arnold proved the case that the Hamiltonian is an analytic function. Moser was able to prove the theorem for the case that the Hamiltonian is 333-times differentiable [162] [163], with the help of Nash implicit function theorem. Another related topic is the Arnold theorem on circle map [10]. It answers the question when a circle map is equivalent to a rotation. Yoccoz [207] was able to prove an if and only if condition for such equivalence, using Brjuno number.

KAM theory for partial differential equations has also been studied [198] [110] [111] [34] [36]. For partial differential equations, KAM theory is studied on a case by case base. There is no general theorem. So far the common types of equations studied are nonlinear wave equations as perturbations of certain linear wave equations, and solition equations under Hamiltonian perturbations. Often the persistent Liouville tori are limited to finite dimensional, sometimes, even one dimensional, i.e. periodic solutions [54].

12.2. Gibbs Measure

Gibbs measure is one of the important concepts in thermodynamic and statistical mechanics. In an effort to understand the statistical mechanics of nonlinear wave equations, Gibbs measure was introduced [155] [33] [35], which is built upon the Hamiltonians of such systems. The calculation of a Gibbs measure is similar to that in quantum field theory. In terms of classical analysis, the Gibbs measure is not well-defined. One of the central questions is which space such Gibbs measure is supported upon. Bourgain [33] [35] was able to give a brilliant answer. For example, for periodic nonlinear Schrödinger equation, it is supported on $H^{-1/2}$ [33]. For the thermodynamic formalism of infinite dynamical systems, Gibbs measure should be an important concept in the future.

12.3. Inertial Manifolds and Global Attractors

Global attractor is a concept for dissipative systems, and inertial manifold is a concept for strongly dissipative systems. A global attractor is a set in the phase space, that attracts all the big balls to it as time approaches infinity. An inertial manifold is an invariant manifold that attracts its neighborhood exponentially. A global attractor can be just a point, and often it is just a set. There is no manifold structure with it. Based upon the idea of reducing the complex infinite dimensional flows, like Navier-Stokes flow, to finite dimensional flows, the concept of inertial manifold is introduced. Often inertial manifolds are finite dimensional, have manifold structures, and most importantly attract their neighborhoods exponentially. Therefore, one hopes that the complex infinite dimensional dynamics is slaved by the finite dimensional dynamics on the inertial manifolds. More ambitiously, one hopes that a finite system of ordinary differential equations can be derived to govern the dynamics on the inertial manifold. Under certain spectral gap conditions, inertial manifolds can be obtained for many evolution equations [46] [48]. Usually, global attractors can be established rather easily. Unfortunately, inertial manifolds for either 2D or 3D Navier-Stokes equations have not been established.

12.4. Zero-Dispersion Limit

Take the KdV equation as an example

$$u_t - 6uu_x + \epsilon^2 u_{xxx} = 0 \;,$$

Lax [115] asked the question: what happens to the dynamics as $\epsilon \to 0$, i.e. what is the zero-dispersion limit ? One can view the KdV equation as a singular perturbation of the corresponding inviscid Burgers equation, do the solutions of the KdV equation converge strongly, or weakly, or not at all to those of the Burgers equation ? In a series of three papers [115], Lax and Levermore investigated these questions. It turns out that in the zero-dispersion limit, fast oscillations are generated instead of shocks or multi-valuedness. Certain weak convergences can also established.

In comparison with the singular perturbation studies of soliton equations in previous sections, our interests are focused upon dynamical systems objects like invariant manifolds and homoclinic orbits. In the zero-parameter limit, regularity of invariant manifolds changes [124].

12.5. Zero-Viscosity Limit

Take the viscous Burgers equation as an example

$$u_t + uu_x = \mu u_{xx} \;,$$

through the Cole-Hopf transformation [88] [44], this equation can be transformed into the heat equation which leads to an explicit expression of the solution to the Burgers equation. Hopf [88], Cole [44], and Whitham [200] asked the question on the zero-viscosity limit. It turns out that the limit of a solution to the viscous Burgers equation can form shocks instead of the multi-valuedness of the solution to the inviscid Burgers equation

$$u_t + uu_x = 0 \;.$$

That is, strong convergence does not happen. Nevertheless, the location of the shock is determined by the multi-valued portion of the solution to the inviscid Burgers equation.

One can also ask the zero-viscosity limit question for Navier-Stokes equations. In fact, this question is the core of the studies on fully developed turbulence. In most of Kato's papers on fluids [99] [100] [101], he studied the zero-viscosity limits of the solutions to Navier-Stokes equations in finite (or small) time interval. It turns out that strong convergence can be established for 2D in finite time interval [101], and 3D in small time interval [100]. Order $\sqrt{\nu}$ rate of convergence was obtained [99]. More recently, Constantin and Wu [49] had investigated the zero-viscosity limit problem for vortex patches.

12.6. Finite Time Blowup

Finite time blowup and its general negative Hamiltonian criterion for nonlinear Schrödinger equations have been well-known [195]. The criterion was obtained from a variance relation found by Zakharov [195]. Extension of such criterion to nonlinear Schrödinger equations under periodic boundary condition, is also obtained [102]. Extension of such criterion to Davey-Stewartson type equations is also obtained [73].

An explicit finite time blowup solution to the integrable Davey-Stewartson II equation, was obtained by Ozawa [169] by inventing an extra conservation law due to a symmetry. This shows that integrability and finite time blowup are compatible. The explicit solution $q(t)$ has the property that

$$\|q(t)\|_{L^2} = 2\sqrt{\pi} \ , \quad \forall t \ , \quad q(t) \notin H^1(R^2) \ , \quad \forall t \ .$$

$$q(t) \in H^s(R^2) \ , \quad s \in (0,1) \ , \quad t \in [0,T] \ ,$$

for some $T > 0$. When $t \to T$,

$$\|q(t)\|_{H^s} \geq C|T - t|^{-s} \to \infty \ , \quad s \in (0,1) \ .$$

The question of global well-posedness of Davey-Stewartson II equation in $H^s(R^2)$ $(s > 1)$ is open.

Unlike the success in nonlinear wave equations, the search for finite time blowup solutions for 3D Euler equations and other equations of fluids has not been successful. The well-known result is the Beale-Kato-Majda necessary condition [21]. There are results on non-existence of finite time blowup [52] [53].

12.7. Slow Collapse

Since 1960's, Yudovich [209] had been promoting the idea of slow collapse. That is, although there is no finite time blowup, if the function's derivative grows to infinity in norm as time approaches infinity, then the function itself should gain randomness in space.

The recent works of Fefferman and Cordoba [52] [53] are in resonance with the slow collapse idea. Indeed, they found the temporal growth can be as fast as e^{e^t} and beyond.

12.8. Burgers Equation

Burgers equation

$$u_t + uu_x = \mu u_{xx} + f(t, x) \ ,$$

was introduced by J. M. Burgers [37] as a simple model of turbulence. By 1950, E. Hopf [88] had conducted serious mathematical study on the Burgers equation

$$u_t + uu_x = \mu u_{xx} \ ,$$

and found interesting mathematical structures of this equation including the so-called Cole-Hopf transformation [88] [44] and Legendre transform. Moreover, Hopf investigated the limits $\mu \to 0$ and $t \to \infty$. It turns out that the order of taking the two limits is important. Especially with the works of Lax [113] [114], this led to a huge interests of studies on conservation laws for many years. Since 1992, Sinai [189] had led a study on Burgers equation with random data which can be random initial data or random forcing [58].

Another old model introduced by E. Hopf [87] did not catch too much attention. As a turbulence model, the main drawback of Burgers equation is lack of incompressibility condition. In fact, there are studies focusing upon only incompressibility [74]. To remedy this drawback, other models are necessary.

12.9. Other Model Equations

A model to describe the hyperbolic structures in a neighborhood of a fixed point of 2D Euler equation under periodic boundary condition, was introduced in [133] [134] [131] [128]. It is a good model in terms of capturing the linear instability. At special value of a parameter, the explicit expression for the hyperbolic structure can be calculated [133] [131]. One fixed point's unstable manifold is the stable manifold of another fixed point, and vice versa. All are 2D ellipsoidal surfaces. Together, they form a lip shape hyperbolic structure.

Another model is the so-called shell model [97] [181] [96] which model the energy transfer in the spectral space to understand for example Kolmogorov spectra.

12.10. Kolmogorov Spectra and An Old Theory of Hopf

The famous Kolmogorov $-5/3$ law of homogeneous isotropic turbulence [106] [107] [105] still fascinates a lot of researchers even nowadays. On the contrary, a statistical theory of Hopf on turbulence [89] [92] [90] [91] [196] has almost been forgotten. By introducing initial probability in a function space, and realizing the conservation of probability under the Navier-Stokes flow, Hopf derived a functional equation for the characteristic functional of the probability, around 1940, published in [89]. Initially, Hopf tried to find some near Gaussian solution [92], then he knew the work of Kolmogorov on the $-5/3$ law which is far from Gaussian probability, and verified by experiments. So the topic was not pursued much further.

12.11. Onsager Conjecture

In 1949, L. Onsager [167] conjectured that solutions of the incompressible Euler equation with Hölder continuous velocity of order $\nu > 1/3$ conserves the energy, but not necessarily if $\nu \leq 1/3$. In terms of Besov spaces, if the weak solution of Euler equation has certain regularity, it can be proved that energy indeed conserves [60] [47].

12.12. Weak Turbulence

Motivated by a study on 2D Euler equation [**210**], Zakharov led a study on infinte dimensional Hamiltonian systems in the spectral space [**211**]. Under near Gaussian and small amplitude assumptions, Zakharov heuristically gave a closure relation for the averaged equation. He also heuristically found some stationary solution to the averaged equation which leads to power-type energy spectra all of which he called Kolmogorov spectra. One of the mathematical manipulations he often used is the canonical transformation, based upon which he classified the kinetic Hamiltonian systems into 3-wave and 4-wave resonant systems.

12.13. Renormalization Idea

Renormalization group approach has been very successful in proving universal property, especially Feigenbaum constants, of one dimensional maps [**45**]. Renormalization-group-type idea has also been applied to turbulence [**168**], although not very successful.

12.14. Random Forcing

As mentioned above, there are studies on Burgers equation under random forcing [**58**], and studies on the structures of incompressible vector fields [**74**]. Take any steady solution of 2D Euler equation, it defines a Hamiltonian system with the stream function being the Hamiltonian. There have been studies on such systems under random forcings [**63**] [**62**].

12.15. Strange Attractors and SBR Invariant Measure

Roughly speaking, strange attractors are attractors in which dynamics has sensitive dependence upon initial data. If the Cantor sets proved in previous sections are also attractors, then they will be strange attractors. In view of the obvious fact that often the attractor may contain many different objects rather than only the Cantor set, global strange attractors are the ideal cases. For one dimensional logistic-type maps, and two dimensional Hénon map in the small parameter range such that the 2D map can viewed as a perturbation of a 1D logistic map, there have been proofs on the existence of strange attractors and SBR (Sinai-Bowen-Ruelle) invariant measures [**23**] [**24**] [**25**].

12.16. Arnold Diffusions

The term Arnold diffusion started from the paper [**12**]. This paper is right after Arnold completed the proof on the persistence of KAM tori in [**11**]. If the degree of freedom is 2, the persistent KAM tori are 2 dimensional, and the level set of the perturbed Hamiltonian is 3 dimensional, then the 2 dimensional KAM tori will isolate the level set, and diffusion is impossible. In [**12**], Arnold gave an explicit example to show that diffusion is possible when the degree of freedom is more than 2. Another interesting point is that in [**12**] Arnold also derived an integral expression for a distance measurement, which is the so-called Melnikov integral [**158**]. Arnold's derivation was quite unique. There have been a lot of studies upon Arnold diffusions [**149**], unfortunately, doable examples are not much beyond that of Arnold [**12**]. Sometimes, the diffusion can be very slow as the famous Nekhoroshev's theorem shows [**164**] [**165**].

12.17. Averaging Technique

For systems with fast small oscillations, long time dynamics are governed by their averaged systems. There are quite good estimates for long time deviation of solutions of the averaged systems from those of the original systems [13] [17]. Whether or not the averaging method can be useful in chaos in partial differential equations is still to be seen.

Bibliography

[1] M. J. Ablowitz and P. A. Clarkson. *Solitons, Nonlinear Evolution Equations and Inverse Scattering*. London Math. Soc. Lect. Note Ser. 149, Cambridge Univ. Press, 1991.

[2] M. J. Ablowitz and J. F. Ladik. A Nonlinear Difference Scheme and Inverse Scattering. *Stud. Appl. Math.*, 55:213, 1976.

[3] M. J. Ablowitz, Y. Ohta, and A. D. Trubatch. On Discretizations of the Vector Nonlinear Schrödinger Equation. *Phys. Lett. A*, 253, no.5-6:287–304, 1999.

[4] M. J. Ablowitz, Y. Ohta, and A. D. Trubatch. On Integrability and Chaos in Discrete Systems. *Chaos Solitons Fractals*, 11, 1-3:159–169, 2000.

[5] M. J. Ablowitz and H. Segur. *Solitons and the Inverse Scattering Transform*. SIAM, Philadelphia, 1981.

[6] R. Adams. *Sobolev Space*. Academic Press, New York, 1975.

[7] L. Almeida, F. Bethuel, and Y. Guo. A Remark on Instability of the Symmetric Vortices with Large Coupling Constant. *Comm. Pure Appl. Math.*, L:1295–1300, 1997.

[8] R. L. Anderson and N. H. Ibragimov. *Lie-Bäcklund Transformations in Applications*. SIAM, Philadelphia, 1979.

[9] D. V. Anosov. Geodesic Flows on Compact Riemannian Manifolds of Negative Curvature. *Proc. Steklov Inst. Math.*, 90, 1967.

[10] V. I. Arnold. Small Denominators. I. Mapping the Circle onto Itself. *Izv. Akad. Nauk SSSR Ser. Mat.*, 25:21–86, 1961.

[11] V. I. Arnold. Proof of a Theorem of A. N. Kolmogorov on the Preservation of Conditionally Periodic Motions under a Small Perturbation of the Hamiltonian. *Uspehi Mat. Nauk*, 18, no.5(113):13–40, 1963.

[12] V. I. Arnold. Instability of Dynamical Systems with Many Degrees of Freedom. *Sov. Math. Dokl.*, 5 No. 3:581–585, 1964.

[13] V. I. Arnold. Applicability Conditions and an Error Bound for the Averaging Method for Systems in the Process of Evolution Through a Resonance. *Dokl. Akad. Nauk SSSR*, 161:9–12, 1965.

[14] V. I. Arnold. Conditions for Nonlinear Stability of Stationary Plane Curvilinear Flows of an Ideal Fluid. *Sov. Math. Dokl.*, 6:773–776, 1965.

[15] V. I. Arnold. Sur la Geometrie Differentielle des Groupes de Lie de Dimension Infinie et ses Applications a L'hydrodynamique des Fluides Parfaits. *Ann. Inst. Fourier, Grenoble*, 16,1:319–361, 1966.

[16] V. I. Arnold. On an Apriori Estimate in the Theory of Hydrodynamical Stability. *Amer. Math. Soc. Transl., Series 2*, 79:267–269, 1969.

[17] V. I. Arnold. *Mathematical Methods of Classical Mechanics*. Springer-Verlag, 1980.

[18] P. Bates, K. Lu, and C. Zeng. Existence and Persistence of Invariant Manifolds for Semiflows in Banach Space. *Mem. Amer. Math. Soc.*, 135, no.645, 1998.

[19] P. Bates, K. Lu, and C. Zeng. Persistence of Overflowing Manifolds for Semiflow. *Comm. Pure Appl. Math.*, 52, no.8:983–1046, 1999.

[20] P. Bates, K. Lu, and C. Zeng. Invariant Foliations Near Normally Hyperbolic Invariant Manifolds for Semiflows. *Mem. Amer. Math. Soc.*, 352, no.10:4641–4676, 2000.

[21] J. T. Beale, T. Kato, and A. Majda. Remarks on the Breakdown of Smooth Solutions for the 3-D Euler Equations. *Comm. Math. Phys.*, 94, no.1:61–66, 1984.

[22] L. Belenkaya, S. Friedlander, and V. Yudovich. The Unstable Spectrum of Oscillating Shear Flows. *SIAM J. Appl. Math.*, 59, no.5:1701–1715, 1999.

[23] M. Benedicks and L. Carleson. The dynamics of the Henon map. *Ann. of Math.*, 133, no.1:73–169, 1991.

[24] M. Benedicks and L.-S. Young. Absolutely Continuous Invariant Measures and Random Perturbations for Certain One-Dimensional Maps. *Ergodic Theory Dynam. Systems*, 12, no.1:13–37, 1992.

[25] M. Benedicks and L.-S. Young. Sinai-Bowen-Ruelle Measures for Certain Henon Maps. *Invent. Math.*, 112, no.3:541–576, 1993.

[26] T. B. Benjamin. The Stability of Solitary Waves. *Proc. R. Soc. Lond.*, A328:153–183, 1972.

[27] G. D. Birkhoff. Some Theorems on the Motion of Dynamical Systems. *Bull. Soc. Math. France*, 40:305–323, 1912.

[28] B. Birnir. Qualitative Analysis of Radiating Breathers. *Comm. Pure Appl. Math.*, XLVII:103–119, 1994.

[29] B. Birnir, H. Mckean, and A. Weinstein. The Rigidity of Sine-Gordon Breathers. *Comm. Pure Appl. Math.*, XLVII:1043–1051, 1994.

[30] C. M. Blazquez. Transverse Homoclinic Orbits in Periodically Perturbed Parabolic Equations. *Nonlinear Analysis, Theory, Methods and Applications*, 10, no.11:1277–1291, 1986.

[31] J. Bona and J. Wu. Zero Dissipation Limit for Nonlinear Waves. *Preprint*, 1999.

[32] J. Bourgain. Fourier Restriction Phenomena for Certain Lattice Subsets and Applications to Nonlinear Evolution Equations. *Geom. and Funct. Anal.*, 3, No 2:107–156, 209–262, 1993.

[33] J. Bourgain. Periodic Nonlinear Schroedinger Equation and Invariant Measures. *Comm. Math. Phys.*, 166:1–26, 1994.

[34] J. Bourgain. Construction of Approximative and Almost-Periodic Solutions of Perturbed Linear Schroedinger and Wave Equations. *Geom. Func. Anal.*, 6(2):201–230, 1996.

[35] J. Bourgain. Invariant measures for the 2D-defocusing nonlinear Schroedinger equation. *Comm. Math. Phys.*, 176, no.2:421–445, 1996.

[36] J. Bourgain. Quasi-Periodic Solutions of Hamiltonian Perturbations of 2D Linear Schroedinger Equations. *Ann. of Math.*, 148, no.2:363–439, 1998.

[37] J. M. Burgers. Mathematical Examples Illustrating Relations Occuring in the Theory of Turbulent Fluid Motion. *Verh. Nederl. Akad. Wetensch. Afd. Natuurk. Sect. 1.*, 17, no.2:53 pp., 1939.

[38] T. Cazenave. *An Introduction to Nonlinear Schrodinger Equations*. IMUFRJ Rio De Janeiro volume 22, 1989.

[39] S. Childress. A Lax pair of 3D Euler equation. *personal communication*, 2000.

[40] S.-N. Chow and J. K. Hale. *Methods of Bifurcation Theory*. Springer-Verlag, New York-Berlin, 1982.

[41] S.-N. Chow, J. K. Hale, and J. Mallet-Paret. An Example of Bifurcation to Homoclinic Orbits. *J. Differential Equations*, 37, no.3:351–373, 1980.

[42] S.-N. Chow, X.-B. Lin, and K. Lu. Smooth Invariant Foliations in Infinite-Dimensional Spaces. *J. Differential Equations*, 94, no.2:266–291, 1991.

[43] S. N. Chow, X. B. Lin, and K. J. Palmer. A Shadowing Lemma with Applications to Semilinear Parabolic Equations. *SIAM J. Math. Anal.*, 20, no.3:547–557, 1989.

[44] J. Cole. On a Quasi-Linear Parabolic Equation Occurring in Aerodynamics. *Quart. Appl. Math.*, 9:225–236, 1951.

[45] P. Collet, J.-P. Eckmann, and O. E. Lanford. Universal Properties of Maps on an Interval. *Comm. Math. Phys.*, 76, no.3:211–254, 1980.

[46] P. Constantin. A Construction of Inertial Manifolds. *Contemporary Mathematics*, 99, 1989.

[47] P. Constantin, W. E, and E. Titi. Onsager's Conjecture on the Energy Conservation for Solutions of Euler's Equation. *Comm. Math. Phys.*, 165, no.1:207–209, 1994.

[48] P. Constantin, C. Foias, B. Nicolaenko, and R. Temam. *Integral Manifolds and Inertial Manifolds for Dissipative Partial Differential Equations*. Applied Mathematical Sciences, 70. Springer-Verlag, New York, 1989.

[49] P. Constantin and J. Wu. The Inviscid Limit for Non-smooth Vorticity. *Indiana Univ. Math. J.*, 45, no.1:67–81, 1996.

[50] B. Coomes, H. Kocak, and K. Palmer. A Shadowing Theorem for Ordinary Differential Equations. *Z. Angew. Math. Phys.*, 46, no.1:85–106, 1995.

[51] B. Coomes, H. Kocak, and K. J. Palmer. Long Periodic Shadowing. *Numerical Algorithms*, 14:55–78, 1997.

[52] D. Cordoba and C. Fefferman. Behavior of Several Two-Dimensional Fluid Equations in Singular Scenarios. *Proc. Natl. Acad. Sci. USA*, 98, no.8:4311–4312, 2001.

[53] D. Cordoba and C. Fefferman. Growth of Solutions for QG and 2D Euler Equations. *J. Amer. Math. Soc.*, 15, no.3:665–670, 2002.

[54] W. Craig and C. E. Wayne. Newton's Method and Periodic Solutions of Nonlinear Wave Equations. *Comm. Pure Appl. Math.*, 46, no.11:1409–1498, 1993.

[55] B. Deng. Exponential Expansion with Silnikov's Saddle-Focus. *J. Differential Equations*, 82, no.1:156–173, 1989.

[56] B. Deng. On Silnikov's Homoclinic-Saddle-Focus Theorem. *J. Differential Equations*, 102, no.2:305–329, 1993.

[57] T. Dombre et al. Chaotic Streamlines in the ABC Flows. *J. Fluid Mech.*, 167, 353–391, 1986

[58] W. E, K. Khanin, A. Mazel, and Ya. Sinai. Invariant Measures for Burgers Equation with Stochastic Forcing. *Ann. of Math.*, 151, no.3:877–960, 2000.

[59] N. M. Ercolani, M. G. Forest, and D. W. McLaughlin. Geometry of the Modulational Instability, Part III: Homoclinic Orbits for the Periodic Sine- Gordon Equation. *Physica D*, 43:349–84, 1990.

[60] G. L. Eyink. Energy dissipation without viscosity in ideal hydrodynamics. I. Fourier analysis and local energy transfer. *Phys. D*, 78, no.3-4:222–240, 1994.

[61] L. D. Faddeev. On the Stability Theory for Stationary Plane-Parallel Flows of Ideal Fluid. *Kraevye Zadachi Mat. Fiziki (Zapiski Nauchnykh Seminarov LOMI, v.21), Moscow, Leningrad: Nauka*, 5:164–172, 1971.

[62] A. Fannjiang and T. Komorowski. Diffusive and Nondiffusive Limits of Transport in Nonmixing Flows. *SIAM J. Appl. Math.*, 62, no.3:909–923, 2002.

[63] A. Fannjiang and G. Papanicolaou. Convection Enhanced Diffusion for Periodic Flows. *SIAM J. Appl. Math.*, 54, no.2:333–408, 1994.

[64] N. Fenichel. Persistence and Smoothness of Invariant Manifolds for Flows. *Ind. University Math. J.*, 21:193–225, 1971.

[65] N. Fenichel. Asymptotic Stability with Rate Conditions. *Ind. Univ. Math. J.*, 23:1109–1137, 1974.

[66] N. Fenichel. Asymptotic Stability with Rate Conditions II. *Ind. Univ. Math. J.*, 26:81–93, 1977.

[67] N. Fenichel. Geometric Singular Perturbation Theory for Ordinary Differential Equations. *J. Diff. Eqns.*, 31:53–98, 1979.

[68] H. Flaschka and D. W. McLaughlin. Canonically Conjugate Variables for the Korteweg-de Vries Equation and the Toda Lattice with Periodic Boundary Conditions. *Progress of Theoretical Physics*, 55, No. 2:438–465, 1976.

[69] M. G. Forest, D. W. McLaughlin, D. J. Muraki, and O. C. Wright. Nonfocusing Instabilities in Coupled, Integrable Nonlinear Schrödinger pdes. *J. Nonlinear Sci.*, 10, no.3:291–331, 2000.

[70] M. G. Forest, S. P. Sheu, and O. C. Wright. On the Construction of Orbits Homoclinic to Plane Waves in Integrable Coupled Nonlinear Schrödinger System. *Phys. Lett. A*, 266, no.1:24–33, 2000.

[71] J. E. Franke and J. F. Selgrade. Hyperbolicity and Chain Recurrence. *J. Diff. Eq.*, 26:27–36, 1977.

[72] S. Friedlander, W. Strauss, and M. Vishik. Nonlinear Instability in an Ideal Fluid. *Ann. Inst. Henri Poincare*, 14, no.2:187–209, 1997.

[73] J. M. Ghidaglia and J. C. Saut. On the Initial Value Problem for the Davey-Stewartson Systems. *Nonlinearity*, 3:475–506, 1990.

[74] M. Ghil, T. Ma, and S. Wang. Structural Bifurcation of 2-D Incompressible Flows. *Indiana Univ. Math. Journal*, 50, no.1:159–180, 2001.

[75] M. Grillakis, J. Shatah, and W. Strauss. Stability Theory of Solitary Waves in the Presence of Symmetry, I. *J. Funct. Anal.*, 74:160–197, 1987.

[76] J. Guckenheimer and P. J. Holmes. *Nonlinear Oscillations, Dynamical Systems, and Bifurcations of Vector Fields.* Springer-Verlag, New York, 1983.

[77] Y. Guo. Instability of Symmetric Vortices with Large Charge and Coupling constant. *Comm. Pure Appl. Math.*, XLIX:1051–1080, 1996.

[78] J. Hadamard. Sur Literation et les Solution Asymptotiques des Equations Differentielles. *Bull. Soc. Math. France*, 29, 1901.

[79] J. K. Hale and X. B. Lin. Symbolic Dynamics and Nonlinear Semiflows. *Ann. Mat. Pura Appl.*, 144:229–259, 1986.

[80] P. Hartman. On the Local Linearization of Differential Equations. *Proc. Amer. Math. Soc.*, 14:568–573, 1963.

[81] A. Hasegawa and Y. Kodama. *Solitons in Optical Communications*. Academic Press, San Diego, 1995.

[82] H. Hasimoto. A Soliton on a Vortex Filament. *J. Fluid Mech.*, 51 (3):477–485, 1972.

[83] D. Henry. Exponential Dichotomies, the Shadowing Lemma and Homoclinic Orbits in Banach Spaces. *Resenhas*, 1, no.4:381–401, 1994.

[84] M. W. Hirsch, C. C. Pugh, and M. Shub. *Invariant Manifolds*. Springer-Verlag, New York, 1977.

[85] D. D. Holm, J. E. Marsden, T. Ratiu, and A. Weinstein. Nonlinear Stability of Fluid and Plasma Equilibria. *Physics Reports*, 123:1–116, 1985.

[86] P. J. Holmes and J. E. Marsden. A Partial Differential Equation with Infinitely Many Periodic Orbits: Chaotic Oscillations of a Forced Beam. *Arch. Rat. Mech. Anal.*, 76:135–166, 1981.

[87] E. Hopf. A Mathematical Example Displaying Features of Turbulence. *Comm. Pure Appl. Math.*, 1:303–322, 1948.

[88] E. Hopf. The Partial Differential Equation $u_t + uu_x = \mu u_{xx}$. *Comm. Pure Appl. Math.*, 3:201–230, 1950.

[89] E. Hopf. Statistical Hydromechanics and Functional Calculus. *J. Rational Mech. Anal.*, 1:87–123, 1952.

[90] E. Hopf. On the Application of Functional Calculus to the Statistical Theory of Turbulence. *Applied probability, Proceedings of Symposia in Applied Mathematics, McGraw-Hill Book Co., for the American Mathematical Society, Providence, R. I.*, 7:41–50, 1957.

[91] E. Hopf. Remarks on the Functional-Analytic Approach to Turbulence. *1962 Proc. Sympos. Appl. Math., American Mathematical Society, Providence, R.I.*, 8:157–163, 1962.

[92] E. Hopf and E. W. Titt. On Certain Special Solutions of the ϕ-Equation of Statistical Hydrodynamics. *J. Rational Mech. Anal.*, 2:587–591, 1953.

[93] M. N. Islam. *Ultrafast Fiber Switching Devices and Systems*. Cambridge University Press, New York, 1992.

[94] C. Jones, N. Kopell, and R. Langer. Construction of the FitzHugh-Nagumo Pulse Using Differential Forms. *in Patterns and Dynamics in Reactive Media, H. Swinney, G. Aris, and D. Aronson eds., IMA Volumes in Mathematics and its Applications, Springer-Verlag, New York*, 37, 1991.

[95] C.K.R.T. Jones and N. Kopell. Tracking Invariant Manifolds with Differential Forms in Singularly Perturbed Systems. *J. Diff. Eq.*, 108:64–88, 1994.

[96] L. Kadanoff, D. Lohse, and N. Schoerghofer. Scaling and Linear Response in the GOY Turbulence Model. *Phys. D*, 100, no.1-2:165–186, 1997.

[97] L. Kadanoff, D. Lohse, J. Wang, and R. Benzi. Scaling and Dissipation in the GOY Shell Model. *Phys. Fluids*, 7, no.3:617–629, 1995.

[98] T. Kapitula and B. Sandstede. Stability of Bright Solitary-Waves Solutions to Perturbed Nonlinear Schrödinger Equations. *Physica D*, 124:58–103, 1998.

[99] T. Kato. Nonstationary Flows of Viscous and Ideal Fluids in R^3. *Journal of Functional Analysis*, 9:296–305, 1972.

[100] T. Kato. Quasi-Linear Equations of Evolution, with Applications to Partial Differential Equations. *Lecture Notes in Mathematics*, 448:25–70, 1975.

[101] T. Kato. Remarks on the Euler and Navier-Stokes Equations in R^2. *Proc. Symp. Pure Math.*, Part 2, 45:1–7, 1986.

[102] O. Kavian. A Remark on the Blowing-Up of Solutions to the Cauchy Problem for Nonlinear Schroedinger Equations. *Trans. Amer. Math. Soc.*, 299, no.1:193–203, 1987.

[103] A. Kelley. The Stable, Center-Stable, Center, Center-Unstable, Unstable Manifolds. *J. Diff. Eqns.*, 3:546–570, 1967.

[104] S. Kichsnassamy. Breather Solutions of the Nonlinear Wave Equation. *Comm. Pure Appl. Math.*, 44:789–818, 1991.

[105] A. N. Kolmogoroff. Dissipation of Energy in the Locally Isotropic Turbulence. *C. R. (Doklady) Acad. Sci. URSS (N.S.)*, 32:16–18, 1941.

[106] A. N. Kolmogoroff. The Local Structure of Turbulence in Incompressible Viscous Fluid for Very Large Reynold's Numbers. *C. R. (Doklady) Acad. Sci. URSS (N.S.)*, 30:301–305, 1941.

[107] A. N. Kolmogoroff. On Degeneration of Isotropic Turbulence in an Incompressible Viscous Liquid. *C. R. (Doklady) Acad. Sci. URSS (N.S.)*, 31:538–540, 1941.

[108] G. Kovacic. Dissipative Dynamics of Orbits Homoclinic to a Resonance Band. *Phys Lett A*, 167:143–150, 1992.

[109] G. Kovacic. Hamiltonian Dynamics of Orbits Homoclinic to a Resonance Band. *Phys Lett A*, 167:137–142, 1992.

[110] S. B. Kuksin. *Nearly Integrable Infinite-Dimensional Hamiltonian Systems*. Lecture Notes in Mathematics, 1556. Springer-Verlag, Berlin, 1993.

[111] S. B. Kuksin. A KAM-Theorem for Equations of the Korteweg-de Vries Type. *Rev. Math. Math. Phys.*, 10, no.3, 1998.

[112] Y. Latushkin, Y. Li, and M. Stanislavova. The Spectrum of a Linearized 2D Euler Operator. *Submitted to Studies in Appl. Math., available at: http://www.math.missouri.edu/~cli*, 2003.

[113] P. Lax. Nonlinear Hyperbolic Equations. *Comm. Pure Appl. Math.*, 6:231–258, 1953.

[114] P. Lax. *The Initial Value Problem for Nonlinear Hyperbolic Equations in Two Independent Variables, Contributions to the Theory of Partial Differential Equations*. Annals of Math. Studies, Vol. 33, pp.211-229, Princeton Univ. Press, Princeton, 1954.

[115] P. Lax and C. D. Levermore. The Small Dispersion Limit of the Korteweg-de Vries Equation, Part I, II, III. *Comm. Pure App. Math.*, 36:253–290, 571–593, 809–830, 1983.

[116] J. L.Bona. On the Stability of Solitary Waves. *Proc. R. Soc. Lond.*, A344:363–374, 1975.

[117] J. L.Bona, P. E. Souganidis, and W. A. Strauss. Stability and Instability of Solitary Waves of KdV Type. *Proc. R. Soc. Lond.*, A411:395–412, 1987.

[118] Y. Li. Backlund Transformations and Homoclinic Structures for the NLS Equation. *Phys. Letters A*, 163:181–187, 1992.

[119] Y. Li. Homoclinic Tubes in Nonlinear Schrödinger Equation Under Hamiltonian Perturbations. *Progress of Theoretical Physics*, 101, no.3:559–577, 1999.

[120] Y. Li. Smale Horseshoes and Symbolic Dynamics in Perturbed Nonlinear Schrödinger Equations. *J. Nonlinear Sciences*, 9:363–415, 1999.

[121] Y. Li. Bäcklund-Darboux Transformations and Melnikov Analysis for Davey-Stewartson II Equations. *J. Nonlinear Sci.*, 10:103–131, 2000.

[122] Y. Li. On 2D Euler Equations: Part I. On the Energy-Casimir Stabilities and The Spectra for Linearized 2D Euler Equations. *J. Math. Phys.*, 41, No.2:728–758, 2000.

[123] Y. Li. A Lax Pair for the Two Dimensional Euler Equation. *J. Math. Phys.*, 42, No.8:3552–3553, 2001.

[124] Y. Li. Persistent Homoclinic Orbits for Nonlinear Schödinger Equation Under Singular Perturbation. *In Press, International Journal of Mathematics and Mathematical Sciences*, 2003.

[125] Y. Li. Chaos and Shadowing Lemma for Autonomous Systems of Infinite Dimensions. *In Press, J. Dynamics and Differential Equations*, 2003.

[126] Y. Li. Chaos and Shadowing Around a Homoclinic Tube. *In Press, Abstract and Applied Analysis*, 2003.

[127] Y. Li. Homoclinic Tubes and Chaos in Perturbed Sine-Gordon Equation. *In Press, Chaos, Solitons and Fractals*, 2003.

[128] Y. Li. Chaos in Partial Differential Equations. *Contemporary Mathematics*, 301: 93–115, 2002.

[129] Y. Li. Existence of Chaos for a Singularly Perturbed NLS Equation. *In Press, International Journal of Mathematics and Mathematical Sciences*, 2003.

[130] Y. Li. Homoclinic Tubes in Discrete Nonlinear Schroedinger Equation Under Hamiltonian Perturbations. *Nonlinear Dynamics*, 31, No.4: 393–434, 2003.

[131] Y. Li. Integrable Structures for 2D Euler Equations of Incompressible Inviscid Fluids. *Proc. Institute Of Mathematics of NAS of Ukraine*, 43, part 1:332–338, 2002.

[132] Y. Li. Melnikov Analysis for Singularly Perturbed DSII Equation. *Submitted, available at: http://xxx.lanl.gov/abs/math.AP/0206272, or http://www.math.missouri.edu/~cli*, 2003.

[133] Y. Li. On 2D Euler equations: II. Lax Pairs and Homoclinic Structures. *Communications on Applied Nonlinear Analysis*, 10, no.1: 1–43, 2003.

[134] Y. Li. Euler Equations of Inviscid Fluids. *Submitted to Advances in Mathematical Sciences, Nova Science Publishers, available at: http://www.math.missouri.edu/~cli*, 2003.

[135] Y. Li. Singularly Perturbed Vector and Scalar Nonlinear Schroedinger Equations with Persistent Homoclinic Orbits. *Studies in Applied Mathematics*, 109:19–38, 2002.

[136] Y. Li, D. McLaughlin, J. Shatah, and S. Wiggins. Persistent Homoclinic Orbits for a Perturbed Nonlinear Schrödinger equation. *Comm. Pure Appl. Math.*, XLIX:1175–1255, 1996.

[137] Y. Li and D. W. McLaughlin. Morse and Melnikov Functions for NLS Pdes. *Comm. Math. Phys.*, 162:175–214, 1994.

[138] Y. Li and D. W. McLaughlin. Homoclinic Orbits and Chaos in Perturbed Discrete NLS System. Part I Homoclinic Orbits. *Journal of Nonlinear Sciences*, 7:211–269, 1997.

[139] Y. Li and S. Wiggins. Homoclinic Orbits and Chaos in Perturbed Discrete NLS System. Part II Symbolic Dynamics. *Journal of Nonlinear Sciences*, 7:315–370, 1997.

[140] Y. Li and S. Wiggins. *Invariant Manifolds and Fibrations for Perturbed Nonlinear Schrödinger Equations*, volume 128. Springer-Verlag, Applied Mathematical Sciences, 1997.

[141] Y. Li and A. Yurov. Lax Pairs and Darboux Transformations for Euler Equations. *Studies in Appl. Math.*, 111: 101–113, 2003.

[142] A. M. Liapunov. *Problème Général de la Stabilité du Mouvement*. Annals of Math. Studies, Vol. 17, Princeton Univ. Press, Princeton, 1949.

[143] V. X. Liu. An Example of Instability for the Navier-Stokes Equations on the 2-Dimensional Torus. *Comm. Partial Differential Equations*, 17, no.11-12:1995–2012, 1992.

[144] V. X. Liu. Instability for the Navier-Stokes Equations on the 2-Dimensional Torus and a Lower Bound for the Hausdorff Dimension of Their Global Attractors. *Comm. Math. Phys.*, 147, no.2:217–230, 1992.

[145] V. X. Liu. A Sharp Lower Bound for the Hausdorff Dimension of the Global Attractors of the 2D Navier-Stokes Equations. *Comm. Math. Phys.*, 158, no.2:327–339, 1993.

[146] V. X. Liu. Remarks on the Navier-Stokes Equations on the Two- and Three-Dimensional Torus. *Comm. Partial Differential Equations*, 19, no.5-6:873–900, 1994.

[147] V. X. Liu. On Unstable and Neutral Spectra of Incompressible Inviscid and Viscid Fluids on the 2D Torus. *Quart. Appl. Math.*, 53, no.3:465–486, 1995.

[148] Y. Liu. Instability of Solutions for Generalized Boussinesq Equations. *Ph. D. Thesis, Brown University*, 1994.

[149] P. Lochak. Arnold Diffusion, a Compendium of Remarks and Questions. *Hamiltonian Systems with Three or More Degrees of Freedom (S'Agaro, 1995), NATO Adv. Sci. Inst. Ser. C Math. Phys. Sci., Kluwer Acad. Publ., Dordrecht*, 533:168–183, 1999.

[150] S. V. Manakov. On the Theory of Two-Dimensional Stationary Self-Focusing of Electromagnetic Waves. *Soviet Physics JETP*, 38, no.2:248–253, 1974.

[151] J. E. Marsden. *Lectures on Mechanics, Lond. Math. Soc. Lect. Note, Ser. 174*. Cambridge Univ. Press, 1992.

[152] V. B. Matveev and M. A. Salle. *Darboux Transformations and Solitons*, volume 5. Springer Series in Nonlinear Dynamics, 1991.

[153] H. P. Mckean and E. Trubowitz. Hill's Operator and Hyperelliptic Function Theory in the Presence of Infinitely Many Branch Points. *Comm. Pure. App. Math.*, 29:143–226, 1976.

[154] H. P. Mckean and P. van Moerbeke. The Spectrum of Hill's Equation. *Inventiones Math.*, 30:217–274, 1975.

[155] H. P. McKean and K. L. Vaninsky. Statistical Mechanics of Nonlinear Wave Equations. *Appl. Math. Sci., 100, Springer, New York*, 100:239–264, 1994.

[156] D. McLaughlin and A. Scott. Perturbation Analysis of Fluxon Dynamics. *Phys. Rev. A*, 18:1652–1680, 1978.

[157] D.W. McLaughlin and E. A. Overman. Whiskered Tori for Integrable Pdes and Chaotic Behavior in Near Integrable Pdes. *Surveys in Appl Math 1*, 1995.

[158] V. K. Melnikov. On the Stability of the Center for Time Periodic Perturbations. *Trans. Moscow Math. Soc.*, 12:1–57, 1963.

[159] C. R. Menyuk. Nonlinear Pulse Propagation in Birefringent Optical Fibers. *IEEE J. Quantum Electron*, 23, no.2:174–176, 1987.

[160] C. R. Menyuk. Pulse Propagation in an Elliptically Birefringent Kerr Medium. *IEEE J. Quantum Electron*, 25, no.12:2674–2682, 1989.

[161] L. D. Meshalkin and Ia. G. Sinai. Investigation of the Stability of a Stationary Solution of a System of Equations for the Plane Movement of an Incompressible Viscous Liquid. *J. Appl. Math. Mech. (PMM)*, 25:1140–1143, 1961.

[162] J. Moser. A Rapidly Convergent Iteration Method and Non-linear Partial Differential Equations. I. *Ann. Scuola Norm. Sup. Pisa*, 20:265–315, 1966.

[163] J. Moser. A Rapidly Convergent Iteration Method and Non-linear Partial Differential Equations. II. *Ann. Scuola Norm. Sup. Pisa*, 20:499–535, 1966.

[164] N. N. Nekhoroshev. An Exponential Estimate of the Time of Stability of Nearly Integrable Hamiltonian Systems. *Uspehi Mat. Nauk*, 32, no.6:5–66, 1977.

[165] N. N. Nekhoroshev. An Exponential Estimate of the Time of Stability of Nearly Integrable Hamiltonian Systems II. *Trudy Sem. Petrovsk*, 5:5–50, 1979.

[166] N. V. Nikolenko. The Method of Poincare Normal Forms in Problems of Integrability of Equations of Evolution Type. *Russian Math. Surveys*, 41:5:63–114, 1986.

[167] L. Onsager. Statistical Hydrodynamics. *Nuovo Cimento*, 6, Supplemento, no.2:279–287, 1949.

[168] S. A. Orszag and V. Yakhot. Renormalization Group Analysis of Turbulence. *Proceedings of the International Congress of Mathematicians, (Berkeley, Calif., 1986), Amer. Math. Soc., Providence, RI*, 1,2:1395–1399, 1987.

[169] T. Ozawa. Exact Blow-Up Solutions to the Cauchy Problem for the Davey-Stewartson Systems. *Proc. R. Soc. Lond.*, 436:345–349, 1992.

[170] J. Palis. A Note on the Inclination Lemma (λ-Lemma) and Feigenbaum's Rate of Approach. *Lect. Notes in Math.*, 1007:630–635, 1983.

[171] J. Palis and W. de Melo. *Geometric Theory of Dynamical Systems.* Springer-Verlag, 1982.

[172] K. Palmer. Exponential Dichotomies and Transversal Homoclinic Points. *J. Differential Equations*, 55, no.2:225–256, 1984.

[173] K. Palmer. Exponential Dichotomies, the Shadowing Lemma and Transversal Homoclinic Points. *Dynamics Reported*, 1:265–306, 1988.

[174] K. Palmer. Shadowing and Silnikov Chaos. *Nonlinear Anal.*, 27, no.9:1075–1093, 1996.

[175] R. L. Pego and M. I. Weinstein. Eigenvalues, and Instabilities of Solitary Waves. *Phil. Trans. R. Soc. Lond.*, A340:47–94, 1992.

[176] O. Perron. Die Stabilitatsfrage bei Differentialgleichungssysteme. *Math. Zeit.*, 32:703–728, 1930.

[177] H. Poincaré. *Les Methodes Nouvelles de la Mecanique Celeste, Vols. 1-3. English Translation: New Methods of Celestial Mechanics. Vols. 1-3, edited by Daniel L. Goroff, American Institute of Physics, New York, 1993.* Gauthier-Villars, Paris, 1899.

[178] J. Poschel and E. Trubowitz. *Inverse Spectral Theory.* Academic Press, 1987.

[179] C. Rogers and W. F. Shadwick. *Bäcklund Transformations and Their Applications.* Academic Press, 1982.

[180] D. H. Sattinger and V. D. Zurkowski. Gauge Theory of Bäcklund Transformations. *Physica D*, 26:225–250, 1987.

[181] N. Schoerghofer, L. Kadanoff, and D. Lohse. How the Viscous Subrange Determines Inertial Range Properties in Turbulence Shell Models. *Phys. D*, 88, no.1:40–54, 1995.

[182] G. R. Sell. Smooth Linearization Near a Fixed Point. *Amer. J. Math.*, 107, no.5:1035–1091, 1985.

[183] J. Shatah and C. Zeng. Homoclinic Orbits for the Perturbed Sine-Gordon Equation. *Comm. Pure Appl. Math.*, 53, no.3:283–299, 2000.

[184] L. P. Silnikov. A Case of the Existence of a Countable Number of Periodic Motions. *Soviet Math. Doklady*, 6:163–166, 1965.

[185] L. P. Silnikov. The Existence of a Denumerable Set of Periodic Motions in Four-dimensional Space in an Extended Neighborhood of a Saddle-Focus. *Soviet Math. Doklady*, 8:54–58, 1967.

[186] L. P. Silnikov. On a Poincare-Birkoff Problem. *Math. USSR Sb.*, 3:353–371, 1967.

[187] L. P. Silnikov. Structure of the Neighborhood of a Homoclinic Tube of an Invariant Torus. *Soviet Math. Dokl.*, 9, 1968.

[188] L. P. Silnikov. A Contribution to the Problem of the Structure of an Extended Neighborhood of a Rough Equilibrium State of Saddle-focus Type. *Math. USSR Sb.*, 10:91–102, 1970.

[189] Ya. Sinai. Statistics of Shocks in Solutions of Inviscid Burgers Equation. *Comm. Math. Phys.*, 148, no.3:601–621, 1992.

[190] S. Smale. A Structurally Stable Differentiable Homeomorphism with an Infinite Number of Periodic Points. *Qualitative Methods in the Theory of Non-linear Vibrations (Proc. Internat. Sympos. Non-linear Vibrations)*, II:365–366, 1961.

[191] S. Smale. Diffeomorphisms with Many Periodic Points. *Differential and Combinatorial Topology (A Symposium in Honor of Marston Morse), Princeton Univ. Press, Princeton, N.J.*, pages 63–80, 1965.

[192] S. Smale. Differentiable Dynamical Systems. *Bull. Amer. Math. Soc.*, 73:747–817, 1967.

[193] H. Steinlein and H. Walther. Hyperbolic Sets and Shadowing for Noninvertible Maps. *Advanced Topics in the Theory of Dynamical Systems, Academic Press, Boston, MA*, pages 219–234, 1989.

[194] H. Steinlein and H. Walther. Hyperbolic Sets, Transversal Homoclinic Trajectories, and Symbolic Dynamics for c^1-Maps in Banach Spaces. *J. Dynamics Differential Equations*, 2, no.3:325–365, 1990.

[195] C. Sulem and P.-L. Sulem. *The Nonlinear Schroedinger Equation, self-focusing and wave collapse*. Applied Mathematical Sciences, 139, Springer-Verlag, New York, 1999.

[196] M. I. Vishik and A. V. Fursikov. *Mathematical Problems of Statistical Hydromechanics*. Akademische Verlagsgesellschaft Geest and Portig K.-G., Leipzig, 1986.

[197] H.-O. Walther. Inclination Lemmas with Dominated Convergence. *ZAMP*, pages 327–337, 1987.

[198] C. E. Wayne. The KAM Theory of Systems with Short Range Interactions. I, II. *Comm. Math. Phys.*, 96, no. 3:311–329, 331–344, 1984.

[199] M. I. Weinstein. Lyapunov Stability of Ground States of Nonlinear Dispersive Evolution Equations. *Comm. Pure Appl. Math.*, 39:51–68, 1986.

[200] G. B. Whitham. *Linear and Nonlinear Waves*. John Wiley and Sons, 1974.

[201] S. Wiggins. *Global Bifurcations and Chaos: Analytical Methods*. Springer-Verlag, New York, 1988.

[202] D. Wirosoetisno and T. G. Shepherd. On the Existence of Two-Dimensional Euler Flows Satisfying Energy-Casimir Stability Criteria. *Phys.Fluids*, 12:727–730, 2000.

[203] O. C. Wright and M. G. Forest. On the Bäcklund-Gauge Transformation and Homoclinic Orbits of a Coupled Nonlinear Schrödinger System. *Phys. D*, 141, no.1-2:104–116, 2000.

[204] J. Wu. The Inviscid Limits for Individual and Statistical Solutions of the Navier-Stokes Equations. *Ph.D. Thesis, Chicago University*, 1996.

[205] J. Wu. The Inviscid Limit of the Complex Ginzburg-Landau Equation. *J. Diff. Equations*, 142, No.2:413–433, 1998.

[206] J. Yang and Y. Tan. Fractal Dependence of Vector-Soliton Collisions in Birefringent Fibers. *Phys. Lett. A*, 280:129–138, 2001.

[207] J.-C. Yoccoz. An Introduction to Small Divisors Problems. *From number theory to physics (Les Houches, 1989), Springer, Berlin*, pages 659–679, 1992.

[208] V. I. Yudovich. Example of the Generation of a Secondary Stationary or Periodic Flow When There is Loss of Stability of the Laminar Flow of a Viscous Incompressible Fluid. *J. Appl. Math. Mech. (PMM)*, 29:527–544, 1965.

[209] V. I. Yudovich. On the Loss of Smoothness of the Solutions of Euler Equations and the Inherent Instability of Flows of an Ideal Fluid. *Chaos*, 10, no.3:705–719, 2000.

[210] V. E. Zakharov. On the Algebra of Integrals of Motion in Two-Dimensional Hydrodynamics in Clebsch Variables. *Funct. Anal. Appl.*, 23, no.3:189–196, 1990.

[211] V. E. Zakharov. *Nonlinear Waves and Weak Turbulence*. American Mathematical Society Translations, Series 2, 182. Advances in the Mathematical Sciences, 36. American Mathematical Society, Providence, RI, 1998.

[212] V. E. Zakharov and A. B. Shabat. Exact Theory of Two-dimensional Self-focusing and One-dimensional Self-modulation of Waves in Nonlinear Media. *Sov. Phys. JETP*, 43(1):62–69, 1972.

[213] W. Zeng. Exponential Dichotomies and Transversal Homoclinic Orbits in Degenerate Cases. *J. Dyn. Diff. Eq.*, 7, no.4:521–548, 1995.

[214] W. Zeng. Transversality of Homoclinic Orbits and Exponential Dichotomies for Parabolic Equations. *J. Math. Anal. Appl.*, 216:466–480, 1997.

Index